BOTANICAL RESEARCH AND PRACTICES

ALFALFA

ECOLOGY, PRODUCTION AND DISEASE MANAGEMENT

BOTANICAL RESEARCH AND PRACTICES

Additional books in this series can be found on Nova's website under the Series tab.

Additional e-books in this series can be found on Nova's website under the e-book tab.

BOTANICAL RESEARCH AND PRACTICES

ALFALFA

ECOLOGY, PRODUCTION AND DISEASE MANAGEMENT

MARIE L. HIRSH
EDITOR

New York

For permission to use material from this book please contact us:
Telephone 631-231-7269; Fax 631-231-8175
Web Site: http://www.novapublishers.com

Additional color graphics may be available in the e-book version of this book.

Library of Congress Cataloging-in-Publication Data

Alfalfa : ecology, production and disease management / editor: Marie L. Hirsh.
pages cm
Includes index.
ISBN: 978-1-62948-752-6 (soft cover)
1. Alfalfa. I. Hirsh, Marie L., editor.
SB205.A4A545 2014
633.3'1--dc23

2013046311

Published by Nova Science Publishers, Inc. † New York

CONTENTS

PREFACE

In this book, the authors present current research in the study of the ecology, production and disease management of the Alfalfa plant. Alfalfa appears to be one of the most interesting of the green crops, since it has high protein content and belongs to a group of major forage crops due to this high productive potential and excellent quality of green biomass. It is an important leguminous crop grown in many countries and has several important roles in different aspects of agriculture which include livestock feed and soil conservation. The authors of this book also explore the biological nitrogen fixation, symbiosis, which contributes greatly to the nitrogen nutrition source on earth.

Chapter 1 – Alfalfa appears to be one of the most interesting of the green crops, since it has high protein content, but also contains other molecules such as polysaccharides or pigments, whose valorization could improve the financial attractiveness of the sector. Extraction of these fractions was therefore investigated as a biorefinery concept, using twin-screw extrusion. The capacity of this technique to separate a liquid extract containing high levels of proteins and pigments, from a solid raffinate rich in hemicelluloses and celluloses, was particularly studied.

This work focused mainly on the characterization of the influence of operating conditions, screw profile, amount of water, temperature, leading to the dehydration of the plant matter while giving a high protein recovery in the extract. The configuration of the screw profile is a crucially important parameter, because it defines the mechanical action on the plant matter, grinding, pressing or shearing. The combination of these actions was investigated in order to achieve both better protein extraction yield, and lower residual water content. Changing the order and spaces between the screw elements, the length of the different sections (alfalfa grinding, liquid/solid extraction, liquid/solid separation) in the extruder, modified the composition of the fractions recovered.

The results show that twin-screw extrusion is an efficient tool for protein extraction from the green plant, since over 50 % of these, corresponding to almost all the soluble ones, can be recovered in the liquid extract, thanks to the combination of pressing and shearing actions. The extract also contained large amounts of pigments such as chlorophyll. Thus, purification could lead to the production of valuable fractions containing proteins and pigments, creating high added value products.

Washing cellulose and hemicellulose fibers leads to the production of a solid raffinate with 47% humidity under optimum conditions. Compared to classical industrial drying, alfalfa dehydration does not appear to be sufficient using this technology. However, the fibrous raffinate could be further valorized using another dehydration step, or directly by further lignocellulose transformation such as for biofuel production.

Chapter 2 – Biological nitrogen fixation, especially symbiosis contributes the most of nitrogen nutrition source on earth. Symbiosis has great potential in agriculture application, to reduce chemical fertilizer, to save energy, and to promote sustainable agriculture.

We evaluated the rhizobia-alfalfa-grass (*Sinorhizobium meliloti – Medicago sativa –Bromus inermis*) as model to study the promotion of nitrogen fixation, remission of the ammonia depression, of which the new knowledge will be a solid theoretical principle for rhizobia-legume-grass system application for ancient-old agricultural practices. Detailed works as following: (A) proposed the screening procedures for high effective symbiotic rhizobia in matter of nodulation ability, nitrogen fixation efficiency and competition capability with native rhizobia. Four strains CCBAU 01290, CCBAU 01199, CCBAU 01314, CCBAU 83164 were confirmed in filed testing; (B) the significant variations in symbiotic efficiency between the 4 rhizobia and 3 host plant cultivars. CCBAU 01199 and CCBAU 01290 were first to know as broader-host arrange rhizobia with all 3 alfalfa cultivars. Inoculated rhizobia could also alter the soil born rhizobia nodulation ability and biodiversity; (C) rhizobia-alfalfa-grass interaction system demonstrated the significantly promoting the alfalfa nodules and two crops yield. The high effective strains CCBAU 01199 and CCBAU 01290 in intercropping system minimized the ammonia depression that high nitrogen (NH_4^+) concentration around the root will deactive the nitrogenase to reduce the N_2 to NH_4^+; (D) evaluation the biodiversity of soil-born microorganism community after inoculation in rhizobia-alfalfa-grass. Understood the biodiversity and dynamic change of microorganism community for 6 months after inoculation will not affected soil bacteria permanently.

Chapter 3 – Alfalfa belongs to a group of major forage crops due to its high productive potential and excellent quality of green biomass and grain, as well as a high content of proteins. The cropping practices for alfalfa, its coverage and height, particularly of seed crops, are conducive to formation of specific agroecological conditions that attract establishment and spread of rodent colonies. Optimal living in alfalfa is furthered by rodents' prominent adaptability to various conditions and types of open habitats, their high reproductive potential and polyphagous character.

Feeding is the primary cause of damage done by rodents in alfalfa crops from the time of sowing, when they consume seeds in the field, throughout the vegetation period when they feed on green biomass, until wintertime when they eat roots. Undoubtedly, they cause the most serious impact on seed crops by eating, injuring and collecting mature seeds. Considering the levels of damage in alfalfa fields, species of the genus Microtus (*Microtus arvalis, Microtus californicus*, *Microtus guentheri, Microtus montanus* and *Microtus pennsylvanicus*) are the most prominent pests. In some alfalfa-growing regions, corresponding damage may be caused by *Thomomys bottae*, *Spermophilus beecheyi*, *Meriones unguiculatus* or *Apodemus sp.*

Modern alfalfa cropping practices include ways to reduce economic losses caused by feeding and other activities of rodents to their economically acceptable level. The many existing alternative control measures, including a topsoil cultivation, building of physical barriers, traps or stimulation of predators, mostly fail to produce desirable effects, so that rodenticide treatments remain the most effective and often most economical method of controlling rodents. The specific distribution and ecological characteristics of each rodent species, different cropping practices and legal limitation for treatments influence the manufacturing and application of various active ingredients and formulations intended for control of rodents in alfalfa fields.

This chapter will survey the most harmful rodent pests in alfalfa crops, their biology, characteristic signs of presence and damage that they cause. Besides some non-chemical methods of control, the available commercial rodenticides will be listed, and their application methods and efficacy described.

In: Alfalfa
Editor: Marie L. Hirsh
ISBN: 978-1-62948-752-6

Chapter 1

TWIN-SCREW EXTRUSION TECHNOLOGY: AN INNOVATIVE STEP FOR THE BIOREFINERY OF ALFALFA

D. Colas[1,2], L. Rigal[1,2] and P. Y. Pontalier[1,2,*]
[1]Université de Toulouse; INP; LCA
(Laboratoire de Chimie Agro-Industrielle); ENSIACET,
Toulouse, France
[2]INRA; LCA (Laboratoire de Chimie Agro-Industrielle);
Toulouse, France

ABSTRACT

Alfalfa appears to be one of the most interesting of the green crops, since it has high protein content, but also contains other molecules such as polysaccharides or pigments, whose valorization could improve the financial attractiveness of the sector. Extraction of these fractions was therefore investigated as a biorefinery concept, using twin-screw extrusion. The capacity of this technique to separate a liquid extract containing high levels of proteins and pigments, from a solid raffinate rich in hemicelluloses and celluloses, was particularly studied.

This work focused mainly on the characterization of the influence of operating conditions, screw profile, amount of water, temperature, leading to the dehydration of the plant matter while giving a high protein

[*] Corresponding author.

recovery in the extract. The configuration of the screw profile is a crucially important parameter, because it defines the mechanical action on the plant matter, grinding, pressing or shearing. The combination of these actions was investigated in order to achieve both better protein extraction yield, and lower residual water content. Changing the order and spaces between the screw elements, the length of the different sections (alfalfa grinding, liquid/solid extraction, liquid/solid separation) in the extruder, modified the composition of the fractions recovered.

The results show that twin-screw extrusion is an efficient tool for protein extraction from the green plant, since over 50 % of these, corresponding to almost all the soluble ones, can be recovered in the liquid extract, thanks to the combination of pressing and shearing actions. The extract also contained large amounts of pigments such as chlorophyll. Thus, purification could lead to the production of valuable fractions containing proteins and pigments, creating high added value products.

Washing cellulose and hemicellulose fibers leads to the production of a solid raffinate with 47% humidity under optimum conditions. Compared to classical industrial drying, alfalfa dehydration does not appear to be sufficient using this technology. However, the fibrous raffinate could be further valorized using another dehydration step, or directly by further lignocellulose transformation such as for biofuel production.

1. Introduction: Alfalfa, a Suitable Candidate for a Biorefinery Process

The large amounts of juice produced from the chopping and pressing of alfalfa prior to industrial drying (reduction in the energetic cost of drying), is often considered as waste. However, this green extract has been recognized by the FAO as an effective source of high quality proteins, with important functional properties (Lamsal et al., 2007; Barbeau and Kinsella, 1988; Douillard and De Mathan, 1994) for animal but also human consumption. This is due to the abundant sources, its nutritive value, and the absence of animal cholesterol (Xie et al., 2008).

Thus, alfalfa appears to be one of the most interesting of the green crops, since it has a high protein content, but also contains lignocelluloses and pigments (such as chlorophyll), whose valorization could improve the financial attractiveness of its cultivation (Volenec et al., 2002; Firdaous et al., 2009). Furthermore, competition between food and biofuel production hinders the development of first generation biofuels (Novalin and Zweckmair, 2009) and would promote alfalfa valorization in line with the green biorefinery concept (Kamm and Kamm, 2004a, 2004b; Andersen and Kiel, 2000).

Green plant fractionation can be carried out using twin-screw extrusion by combining chemical, thermal and mechanical actions. This technique has already been used for the extraction of hemicelluloses from lignocellulosic matter such as poplar (N'Diaye et al., 1996) or wheat straw (Zeitoun et al., 2010). Extrusion improves mechanical lysis of the cells and is an efficient tool for conducting several operations using a single device, e.g., i) mixing the matter and the solvent, ii) liquid/solid extraction, and iii), an initial liquid/solid separation step.

2. Twin-Screw Extrusion Design

2.1. Presentation of the Twin-Screw Extruder

Twin-screw extrusion plays an important role in the food industry (e.g., extrusion-cooking of cereals, snacks and pet food production, vegetable protein texturizing, Roustel, 2000). More recently, the range of materials treated with this technique has been widened, with for example biomass fractionation (Rigal, 2000). Thanks to twin-screw extrusion, mechanical, thermal and chemical actions on the material can take place in a single step, and thus extraction of solutes from vegetable matter can be undertaken.

Twin-screw extruder action is mainly due to screw elements with different functions (Figure 1):

- forward pitch screws: conveying action
- monolobe paddle screws: radial compression and shearing action
- bilobe paddle screws: significant mixing and shearing action, conveying and axial compression action in combination with the forward pitch screws
- reverse pitch screws: intensive shearing, mixing and also strong axial compression in combination with the forward pitch screws (Rigal, 2000; N'Diaye et al., 1996)

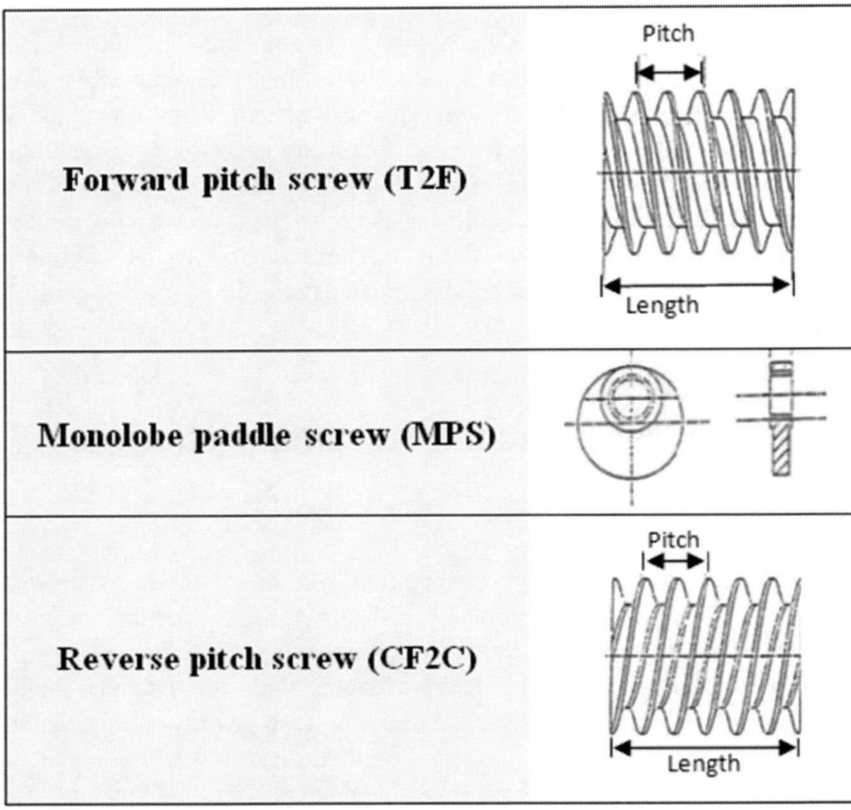

Figure 1. Examples of different screw elements.

The arrangement of the different screw elements in different positions and with different spacing defines the screw 'configuration', which is the main factor influencing extrusion performance (product transformation, residence time distribution, mechanical energy input) (Gogoi et al., 1996; Choudhury et al., 1998; Gautam and Choudhury, 1999a; Gautam and Choudhury, 1999b; Evon et al., 2009).

Experiments were conducted on a co-penetrating and co-rotating twin-screw Clextral BC 45 (Firminy, France) extruder (Figure 2). This is composed of two parallel screw shafts inside a barrel, on which are placed segmental screw elements, each of 50 or 100 mm in length.

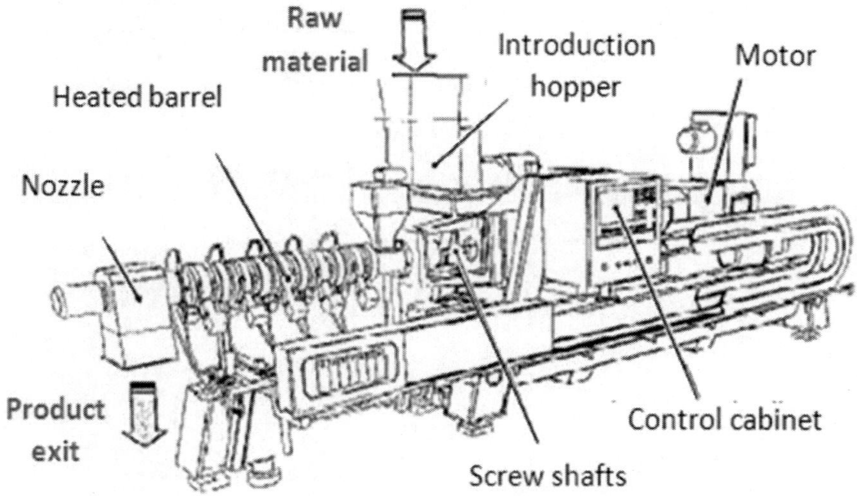

Figure 2. Clextral BC 45 twin-screw extruder.

The extruder barrel consists of seven, 20 cm long modules (Figure 3), with one filtration module inserted at position F (screw profiles 1, 2 and 3), or two filtration modules at positions D and F (screw profile 4). The latter are each composed of four hemispherical dishes with 1 mm diameter perforations, and an outlet on the seventh module.

Modules C, D, E and G are temperature-regulated with heating and cooling systems. The reverse pitch screws in module G (and in module E for screw profile 4) create a counter current, carrying the matter in the opposite direction immediately after the filtration module, producing a very strong flow-restricting element with a high ability to generate pressure.

This action increases extruder filling, leading to the formation of a dynamic plug. The pressed and squeezed plant matter produces a liquid that can be separated in the filtration section, inserted to collect a liquid filtrate and a solid raffinate separately at the exit of the extruder. Barrel temperature and screw rotation speed are monitored from a console.

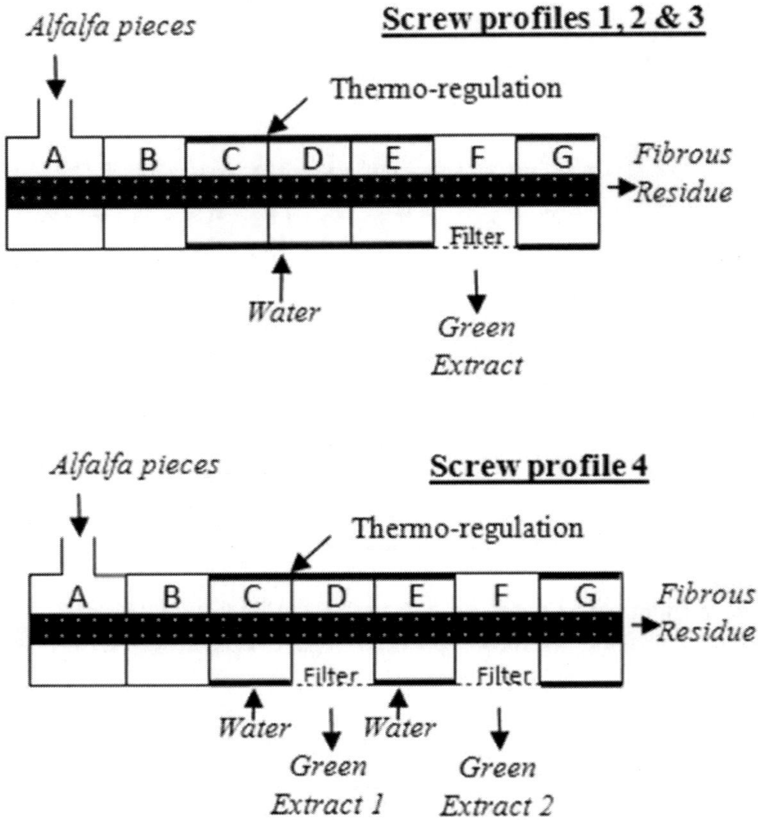

Figure 3. Schematic representation of the twin-screw extruder barrel.

2.2. Description of the Screw Profiles

Chopped frozen alfalfa (5 centimeter pieces) is introduced into the extruder in the first barrel segment, at a mean rate of (10.5 ± 0.9) $kg.h^{-1}$. A pump is used to introduce water at the beginning of the fourth module for screw profiles 1 and 2. Two pumps are used for profile 4, introducing water twice, at the end of module C and at the beginning of module E.

The two different screw profiles tested here were based on previous studies on treatment of plant matter (Evon et al., 2009; Maréchal, 2001; Amalia Kartika et al., 2005; Zeitoun et al., 2010).

2.2.1. First Screw Profile

The first screw profile (Figure 4 (a)) is composed of trapezoidal double-thread screws (T2F) inserted at the beginning, ensuring efficient transport. Conveying, double-thread screws (C2F) mainly exert a conveying action and some mixing of the alfalfa pieces. Bilobe paddle-screws (BB) inserted in the third module apply a pronounced shearing effect and also induce mixing of the matter. Water is added at the beginning of module D, ahead of the reverse pitch screws (CF1T) in module F, which carry the matter in the opposite direction. CF2C are positioned immediately after the filtration module, exerting intensive shearing and guaranteeing the formation of a dynamic plug for green juice expression and separation.

2.2.2. Second Screw Profile

On the second screw profile (Figure 4 (b)), monolobe paddle-screws (MPS) are inserted before the bilobe paddle-screws, in order to exert a higher shearing effect on the plant matter. The contact surface between alfalfa and solvent is thus enhanced.

2.2.3. Screw Profile 3

The third screw profile (Figure 4 (c)) is the same as profile 2, except for the last two modules where reverse pitch screws (C1FTC and CF1T) are closer to the filter, in order to limit water absorption by the alfalfa fibers, after expression of the extract.

2.2.4. Screw Profile 4

This last screw profile has been developed in order to first extract the proteins, and then dehydrate the alfalfa. It is thus composed of two filtration zones, in modules D and F (Figure 4 (d)). CF1C screws are placed after the filter in order to create the dynamic plug. After collection of the green filtrate 1, water is added a second time in order to wash the already pressed fibers. Reverse pitch screws (C1FTC and C1FT) are inserted on the filter, to avoid reabsorption of the liquid pressed from the alfalfa due to the second dynamic plug.

(A) Screw Profile 1

A		B		C			D		E		F		G		
T2F	C2F	C2F	C2F	C2F	BB	C2F	C2F	C2F	C2F	C2F	C2F	CF1T	CF2C	C2F	C2F
66	50	33	33	25	90°	33	33	33	33	25	25	15	-25	33	33

(B) Screw Profile 2

A		B		C				D		E		F		G		
T2F	C2F	C2F	MPS	C2F	BB	C2F	C1F	C1F	BB	C1F	C1F	C1F	CF1C	CF1T	C1FTC	C1F
66	50	33	45 °	25	90°	33	33	25	90°	33	33	33	25	15	-15	25

(C) Screw Profile 3

A		B		C			D			E		F		G		
T2F	C2F	C2F	MPS	C2F	BB	C1F	C1F	C1F	BB	C1F	C1F	C1F	CF1T	C1FTC	C1F	C1F
66	50	33	45 °	25	90°	33	25	25	90°	33	33	33	15	15	33	25

(D) Screw Profile 4																
A		B		C			D			E		F		G		
T2F	C2F	C2F	MPS	C2F	BB	C1F	C1F	C1F	CF1C	C1F	C1F	C1F	CF1T	C1FTC	C1F	C1F
66	50	33	45 °	25	90°	33	25	25	-25	33	33	33	15	-15	33	25

(T2F: trapezoidal double-thread screw; C2F: conveying double-thread screw; C1F: conveying simple screw; BB: bilobe paddle-screw; MPS: monolobe paddle-screw; CF1C: reverse pitch screw; CF1TC: reverse pitch screw. Screw type and screw pitch or length (mm) are represented).

Figure 4. Screw profiles.

3. Feasibility and Study of the Influence of Extrusion Operating Conditions on the Dehydration of Alfalfa and the Production of a Green Extract Rich in Proteins

3.1. Chemical Composition of the Alfalfa

Alfalfa whole plant (Angers, France), frozen directly after harvesting, was used for the trials. Two different batches were received, for the two different parts of the work. The chemical composition of these batches is given below (Table 1).

Table 1. Chemical composition of the alfalfa

	Alfalfa whole plant Batch 1 Trials 1 to 16	**Alfalfa whole plant Batch 2 Trials 17 to 26**
Dry Matter (%)	37.9 ± 0.8	19.9 ± 0.5
Ash (% of the DM)	13.7 ± 1.6	10.2 ± 1.5
Protein (% of the DM)	15.3 ± 1.2	28.4 ± 1.5
Lipids (% of the DM)	8.3 ± 0.3	5.2 ± 0.3
Celluloses (% of the DM)	16.2 ± 1.0	20.3 ± 1.1
Hemicelluloses (% of the DM)	19.5 ± 0.3	15.6 ± 1.7
Lignin (% of the DM)	7.6 ± 1.4	10.3 ± 0.3

Means and variability were calculated from at least three tests.

The dry matter content of the samples was determined according to the French NF V 03-903 standard, at 103 °C for a period of 24 h.

Ash content was determined by thermogravimetric analysis after 3-hour incineration at 550 °C.

Protein content was determined using the Kjeldahl method (NF V 18-100) on an automatic device. The overall amount of nitrogen obtained with this technique is converted into protein proportion using an empirical coefficient of 6.25 (FAO, 2002).

The fiber content (celluloses, hemicelluloses and lignin) was determined using Acid Detergent Fiber - Neutral Detergent Fiber (ADF-NDF method (Van Soest and Wine, 1968a, 1968b)).

The alfalfa was thawed for a day at +4 °C and chopped into five-centimeter pieces, using a Viking plant crusher (STIHL AG & Co, Germany).

3.2. Trial Processing

Trials were first carried out at different screw rotation speeds (from 60 to 140 rpm), in order to determine the operating conditions limits, to achieve the separation of a solid and a liquid fraction with no filtrate reflux in the extruder feed section.

Using the results obtained, two experimental designs, one for each screw profile, were set up in order to test the influence of the extrusion parameters (temperature, water addition) on the quality of alfalfa fractions.

A Doelhert matrix was used (Table 2) to define the experimental conditions for each trial. Barrel temperature could be regulated between 25 °C and 50 °C on four out of the seven extruder modules. The calculated liquid/solid ratio was fixed between 0 (no solvent introduced) and 4 (maximum solvent introduced: four times the weight of alfalfa).

Real liquid/solid ratios were calculated at the end of the trials, by taking the quantified humidity of the whole alfalfa introduced, and the water content of the fractions obtained after extrusion. The differences between the theoretical and real L/S ratio come mainly from pump flow rate variations during the trials.

Extrusion was carried out for at least 20 min before sampling, in order to achieve a steady state with stable flow rate and barrel temperature. For each trial, solid raffinate and green filtrate were then collected over a period of 10 to 15 min.

Outlet parameters studied were:

- Green filtrate and raffinate dry matter, to evaluate the juice expression rate (DM yield (%) = DM_f/DM_a x 100, where DM_a is the quantity of alfalfa dry matter introduced into the extruder for one trial and DM_f is the quantity of dry matter in the filtrate obtained from the trial).
- The protein fraction amount, to analyze the protein extraction rate (Protein yield (%) = P_f/P_a x 100, where P_a is the quantity of alfalfa proteins introduced into the extruder for one trial and P_f is the quantity of proteins in the separated filtrate for the same trial).

Table 2. Doehlert's experimental design

Trial Index Number	X1	Temperature (°C)	X2	Liquid/solid Ratio (calculated)	Real Liquid/solid ratio
4	0	37.5	0	2	4.5
8	0	37.5	0	2	5.4
3	+1	50	0	2	4.7
5	-1	25	0	2	4.4
7	+0.5	43.8	+0.87	3.7	7.8
1	-0.5	31.3	-0.87	0.3	2.0
2	+0.5	43.8	-0.87	0.3	2.1
6	-0.5	31.3	+0.87	3.7	7.7
12	0	37.5	0	2	5.2
16	0	37.5	0	2	5.0
11	+1	50	0	2	5.1
13	-1	25	0	2	4.8
15	+0.5	43.8	+0.87	3.7	8.4
9	-0.5	31.3	-0.87	0.3	2.1
10	+0.5	43.8	-0.87	0.3	2.1
14	-0.5	31.3	+0.87	3.7	7.3

X1 and X2 represent the coding values for the Nemrod software of the barrel temperature and the calculated liquid/solid ratio respectively.

Statistical analysis of the results was made with the New Efficient Methodology for Research using Optimal Design (NEMROD) Software (Mathieu and Phan-Than-Luu, 1992). The coefficients of the polynomial models relating the operating variables to each response studied were calculated from statistical analysis of data using the Nemrod software, which was also used to plot isoresponse curves.

The differences between the experimental values and the predicted results for the statistical analysis of polynomial models are shown in Table 3.

Table 3. Experimental and statistically calculated values for the responses to the experimental designs

	Protein extraction yield		Dry matter yield in filtrate		Protein content in filtrate	
Trial number	**Experi-mental (%)**	**Nemrod software prediction (%)**	**Experimental (%)**	**Nemrod software prediction (%)**	**Experimental (% of the DM)**	**Nemrod software prediction (% of the DM)**
1	40.3	43.1 *-2.8*	51.1	50.1 *+1.0*	23.6	23.5 *+0.1*
2	46.4	43.6 *+2.8*	52.2	53.1 *-0.9*	21.5	21.9 *-0.4*
3	48.5	51.1 *-2.6*	53.7	52.9 *+0.8*	23.5	23.2 *+0.3*
4	51.9	52.2 *-0.3*	51.1	51.2 *-0.1*	25.4	24.1 *+1.3*
5	53.9	51.3 *+2.6*	51.5	52.3 *-0.8*	24.9	25.2 *-0.3*
6	57.0	59.2 *-2.2*	53.3	52.6 *+0.7*	24.2	23.9 *+0.3*
7	60.0	57.8 *+2.2*	48.8	48.8 *-0.0*	23.3	23.4 *-0.1*
8	54.9	54.6 *+0.3*	51.1	51.1 *+0.0*	23.0	24.3 *-1.3*
9	6.4	7.3 *-0.9*	3.6	4.2 *-0.6*	30.3	30.3 *+0.0*

Table 3. (Continued)

	Protein extraction yield		Dry matter yield in filtrate		Protein content in filtrate	
Trial number	**Experi-mental (%)**	**Nemrod software prediction (%)**	**Experimental (%)**	**Nemrod software prediction (%)**	**Experimental (% of the DM)**	**Nemrod software prediction (% of the DM)**
10	9.8	9.2 *+0.6*	6.0	5.6 *+0.4*	29.7	29.7 *+0.0*
11	46.3	47.1 *-0.8*	28.0	28.5 *-0.5*	30.8	30.8 *+0.0*
12	43.8	47.0 *-3.2*	31.1	31.1 *-0.0*	30.4	30.6 *-0.2*
13	36.8	35.9 *+0.9*	20.6	20.6 *+0.0*	28.7	28.7 *-0.0*
14	50.4	51.4 *-1.0*	30.9	31.6 *-0.7*	28.9	28.9 *+0.0*
15	58.7	58.0 *+0.7*	35.8	35.3 *+0.5*	31.4	31.4 *+0.0*
16	49.1	45.4 *+3.7*	33.8	30.2 *+3.6*	30.8	30.6 *+0.2*

Values in italics show the difference between the Nemrod calculation and the experimental value. Equations of the model (as calculated with NEMROD software): First experimental design: Protein extraction yield = $53.3-0.7X_1+7.7X_2-1.1X_1^2-2.2X_2^2-0.8X_1X_2$; Dry matter extraction yield = $51.1-0.1X_1-0.3X_2+1.6X_1^2-0.2X_2^2-3.0X_1X_2$; Protein content in the filtrate = $24.3-1.0X_1+0.5X_2+0.1X_1^2-1.2X_2^2+0.6X_1X_2$; Second experimental design: Protein extraction yield = $47.2+4.7X_1+23.7X_2-3.5X_1^2-16.0X_2^2+3.0X_1X_2$; Dry matter extraction yield = $31.2+3.4X1+14.2X_2-5.31X_1^2-11.8X_2^2+2.1X_1X_2$; Protein content in the filtrate = $30.5+1.2X_1-0.1X_2-0.9X_1^2-0.4X_2^2+1.8X_1X_2$

3.3. Evaluation of Alfalfa Dehydration and Protein Recovery Yield in the Filtrate under Extrusion, with Pressing Conditions

The first screw profile has been defined in order to exert a high-pressure action on the plant matter (using two consecutive counter-current /reverse pitch screws), but with a low shearing effect, limiting cell disruption.

3.3.1. Evaluation of the Dehydration of Alfalfa

Analysis of the isoresponse curves from the model (Nemrod software) (Figure 5) shows the dry matter content of the solid fibrous residue (or cake meal), under the different experimental conditions.

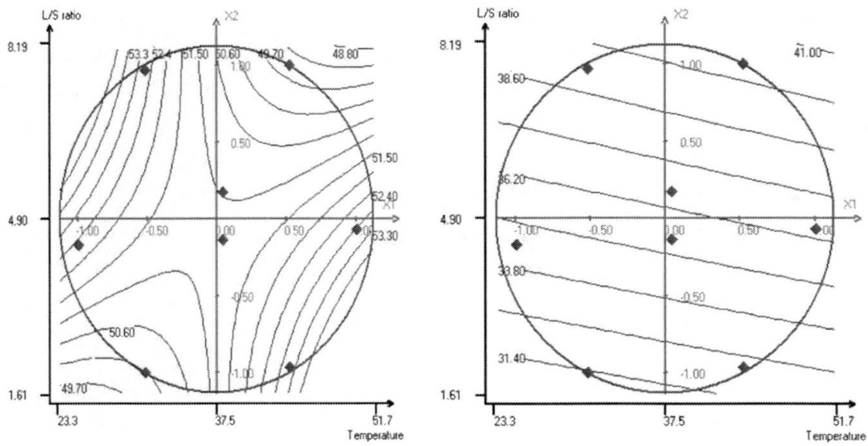

(A) (B)

Figure 5. A) Dry matter content of the solid fibrous residue, according to extrusion temperature and liquid/solid ratio, for screw profile 1 (alfalfa DM before extrusion = 37.9 %); B) Dry matter recovery yield in the liquid extract, according to extrusion temperature and liquid/solid ratio, for screw profile 1.

Extrusion allows partial dehydration of the matter for all the trials, even when the most water is introduced. The dry matter content of the solid residues increases from 38 to 53 % under the two optimum conditions shown on Figure 5 (A). For dehydration, these conditions can be found at the extremes of the isoresponse curves: for the highest temperatures and the lowest L/S ratios.

Obtaining a dryer fibrous residue when less water is introduced into the device would seem to be obvious, however the second optimum reached for a high liquid/solid ratio at room temperature, is less so. A large quantity of water appears to produce better 'wringing out' of the alfalfa fibers and consequently, better green juice expression, and this wringing out seems to be more efficient at the lowest temperatures. When the temperature is lower, the viscosity of the plant matter is higher and water expression is better.

When water is added, extruder filling is increased, which seems to favor the dry matter extraction yield (Figure 5 (B)). It can be assumed that when the L/S ratio is high, increasing the temperature improves expression of the extracted solutes because it decreases the viscosity of the extract solution. The viscosity of this solution can be locally high due to the presence of polysaccharides and proteins, extracted when the shearing effect of the filter is higher. However, under these conditions, plant squeezing is less efficient, because the matter is too fluid, and therefore the water content of the raffinate is higher. At the lowest temperatures, the extract has a lower solute concentration. Therefore, it appears difficult to have an optimum condition for both plant dehydration and recovery of soluble molecules

3.3.2. Efficiency of Protein Extraction

The aim of the extrusion-pressing step was to produce a liquid extract with the highest protein content.

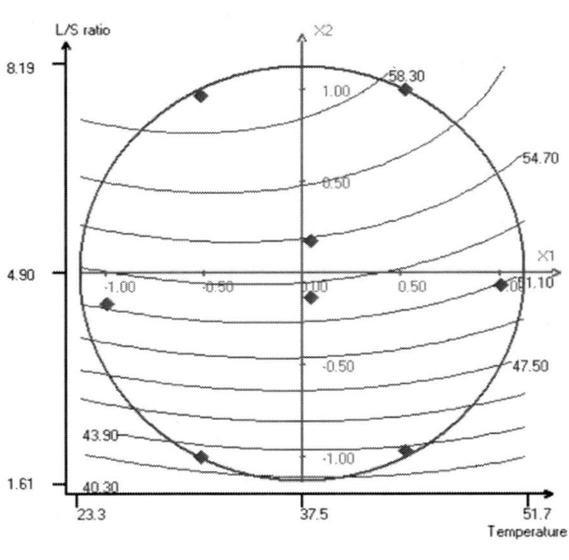

Figure 6. Protein recovery yield in the green filtrate, for screw profile 1.

The highest protein extraction yield in the green extract (percentage of initial proteins recovered in the filtrate; 58.3 % of total proteins introduced) is obtained for the highest liquid/solid ratios (8.1, Figure 6). A reference extraction sample was tested using magnetic stirring of alfalfa at a L/S ratio of 99/1, for 7 h at room temperature. Only 8.3 % of the proteins were extracted in the water, showing the effectiveness of the extrusion process, where more than half of the proteins can be extracted.

Leaching the fibers with high amounts of water leads to an improvement of protein recovery in the liquid extract. For the low L/S ratios, alfalfa pressing leads to the recovery of a less concentrated fluid, consisting mainly of interstitial water, without solutes.

Nevertheless, analysis of the green extract also indicates (Colas et al., 2013) that the mechanical action leads to the extraction of many other molecules, or structures such as cell fragments that are removed from the liquid after centrifugation.

3.4. Evaluation of Alfalfa Dehydration and Protein Recovery Yield in the Filtrate under Extrusion, with Shearing Conditions

In the second screw profile (Figure 4 (b)), extra mixing screw elements (monolobe and bilobe paddle-screws) are included along the barrel to enhance cell disruption and water/plant interaction. Furthermore, the use of C1F mono-thread conveying screws instead of double-thread screws produces better thrust power and a higher shearing effect. Which is why we refer to this screw profile as extrusion under shearing action, compared to screw profile 1.

3.4.1. Evaluation of the Dehydration of Alfalfa

It is difficult to define an optimum condition for alfalfa dehydration, since it is low, with a dry matter content of the raffinate remaining below 40 % wt (Figure 7 (A)).

The highest dehydration within the limits of the experimental design is obtained for the highest L/S ratios when profile 2 is used. The results indicate that the dehydration efficiency changes with experimental conditions. For lower L/S ratios, the solid residue dry matter is maximum both for the lowest and highest temperatures tested. Furthermore, at constant temperature, increasing the L/S ratio from 0.1 to 3.5 has almost no effect on residue dry matter content. For L/S ratios above 3.5, dehydration efficiency increases as well, whatever the temperature.

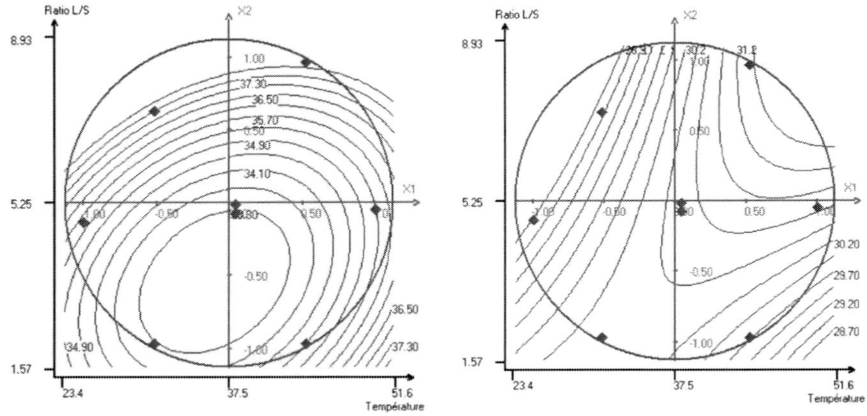

(A) (B)

Figure 7. A) Solid fibrous residue dry matter content (% of whole plant weight), according to extrusion temperature and liquid/solid ratio, for screw profile 2 (alfalfa DM before extrusion = 37.9 %); B) Dry matter extraction yield in the green juice (% of the total DM in the green juice).

The dry matter expression yield (Figure 7 (B)), calculated using dry matter content of the green juice (data not shown), shows a clear optimum condition at 40 °C and L/S ratio of 5. The yield is almost temperature independent for low L/S ratios, but its influence becomes noticeable at L/S ratios above 3.5. Conversely, L/S ratio has a significant influence on dry matter yield for the lowest temperature values. However, it must be stressed that above an L/S ratio of 3.5, it is the temperature that has the most significant effect. Increasing the temperature improves DM recovery, maybe because diffusion of molecules is facilitated. However, squeezing of the plant matter is more difficult to achieve, due to the presence of the fluid, and thus dehydration is hindered.

3.4.2. Efficiency of the Protein Extraction

The best conditions for alfalfa dehydration, determined previously with profile 2, do not match the operating conditions leading to the best protein extraction yield in the green extract (Figure 8).

The L/S ratio greatly influences protein recovery since the yield increases from 6 % wt at L/S = 1.6 to 54 % wt at L/S = 8 at a temperature of 37.5 °C. For all temperatures tested, the best L/S ratio appears to be 8. At low L/S ratios temperature has almost no influence, despite possible denaturation of some proteins above 40 °C, which could hinder their extraction. Nevertheless, at high L/S ratios, increasing the temperature from 23 °C to 50 °C improves protein recovery yield from 44 % wt to 55 % wt. This result could be explained by the decrease in viscosity of the solution, allowing it to flow out of the plant easier. The measurement of the protein content in the extract indicates that a temperature influence has to be considered. The highest protein content in the green juice is 31.4 % of dry matter, for highest ratios and highest temperatures.

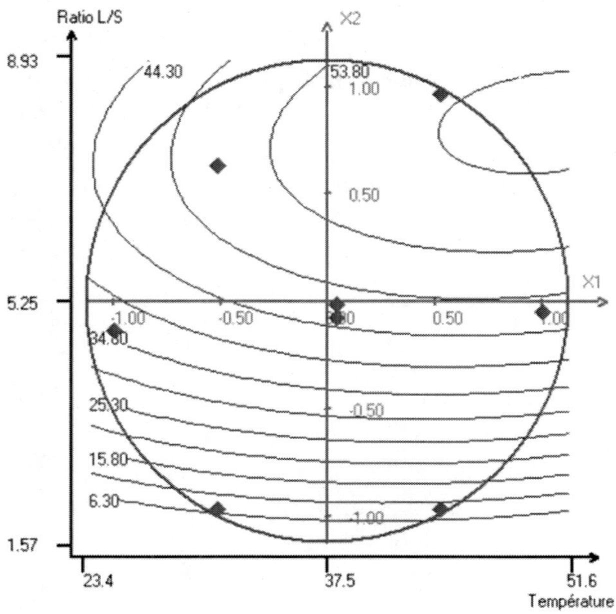

Figure 8. Protein extraction yield in the green juice.

3.5. Comparison of the Two Screw Profiles

High protein recovery yields are obtained with both screw profiles (58 % of the weight of introduced proteins for the first and 52 % wt for the second). However, the protein content of the filtrate is higher with the second screw profile (25 % of the DM are proteins with screw profile 1, and 31.2 % of the DM are proteins for screw profile 2, Colas et al., 2013). As the number of mixing elements is higher in the latter profile, it would seem that cell disruption and fiber washing are both improved, allowing good protein extraction from the alfalfa.

Comparison between the two profiles indicates that the second is less efficient for alfalfa dehydration, with a dry matter content of 38.1 %, lower than the solid residue dry matter content obtained with the first screw profile (55 %). With the second profile, water tends to be absorbed by the plant and remains entrapped in the plant matrix, leading to very low dehydration. Water recovery in the filtrate could be hindered by the structure of the plant matter at the end of the second extruder profile, or by raffinate pressing difficulties in the reverse pitch screws (CF1C).

Hence, the first screw profile is the most efficient for both alfalfa dehydration and protein recovery. Twin-screw extrusion could be a new efficient way to extract proteins from vegetable matter, while washing cellulose fibers for further treatment. In this study on alfalfa, more than 50 % of the global proteins can be released in the liquid extract after extrusion. In addition, the fibrous residue could be further valorized for the usual animal food applications, with the first dehydration step being carried out in the extruder, giving a 50 % DM content.

Twin-screw extruder design is a crucially significant parameter. When mechanical action on plant tissue is stronger, protein recovery is better, although it leads to higher liquid retention. The configuration of the screw profile should be optimized in order to also optimize its action on the matter, to achieve a high protein extraction yield with low residue water content. The variability of the raw material must also be taken into account. The second part of our work thus consisted of an optimization of the screw extruder design, using a different batch of alfalfa.

4. Study of the Influence of Twin-Screw Extruder Configuration on the Dehydration of Alfalfa and Production of a Green Extract Rich in Proteins

The batch of alfalfa used for this part of the work had a different chemical composition (Table 1), it was younger, with a higher water content (80.1 % humidity).

The method for processing the trials was the same as for the first part of this work. However, at the end of some trials, the extruder has been stopped manually and opened, in order to recover the matter remaining on the screws, in each section of the barrel. This was then weighed, and its dry weight calculated. In addition, green filtrates were centrifuged at 4000 g for 20 min at room temperature (Sigma 6K15 centrifuge), in order to eliminate solid particles and obtain a clear brown juice that would be submitted to further treatment in order to separate and purify proteins.

4.1. Screw Profile 2

First of all, twin-screw extrusion trials were carried out with screw profile 2, using the second batch of plant material. The purpose of this investigation was to test the influence of the quality of the alfalfa on extrusion feasibility, and on the quality of the fractions, under precise operating conditions.

The effect of operating conditions (alfalfa inlet flow rate, liquid/solid ratio) on the quality of the filtrates and the solid fibrous residues was then investigated for each screw profile. Trials have been carried out with different liquid/solid (L/S) ratios and solid flow rates (the ratio between the feed rate Q_A and the screw speed). Results are shown in Table 4.

Compared to the results obtained with the first batch of alfalfa, it appears that the solid fibrous residue, or cake meal, produced during trials 17 to 20 (26.3 to 30.1 % of DM) is always wetter than the cake meal obtained when processing the first batch of alfalfa (33.3 to 37.8 % of DM with screw profiles 1 and 2). This shows that younger alfalfa with higher water content, is more difficult to dehydrate, because plant and water mixing creates a viscous matter, which is not easy to press, and the water remains inside the plant. The same observation can be made concerning the protein extraction yield in the filtrate (35.9 % of the proteins recovered under the best conditions with alfalfa batch 2, versus more than 50 % with batch 1). The leaching of the fibers is more difficult with younger tissues, thus proteins remain in the plant.

Table 4. Experimental results for protein extraction and alfalfa dehydration by CLEXTRAL BC 45 twin-screw extruder: Screw Profile 2

Trial Number	**17**	**18**	**19**	**20**
Screw Profile	2	2	2	2
Liquid/solid Ratio	6.6	4.0	7.6	7.1
Alfalfa flow rate Q_A (kg/h)	27.7	11.8	11.8	14.5
Water flow rate Q_W (kg/h)	13.5	0.0	8.4	9.0
Filtrate flow rate Q_F (kg/h)	29.4	5.4	14.7	16.9
Cake meal flow rate Q_{CM} (kg/h)	12.0	6.6	6.0	6.0
Foot content of the filtrate (%wt)	14.1	20.9	12.3	5.3
Dry Matter content of the clarified filtrate (%)	4.6	7.9	4.0	4.3
Dry Matter content of the cake meal (%)	30.1	27.0	26.3	27.6
DM recovery in the clarified filtrate (%)	**20.7**	**14.2**	**21.7**	**23.7**
Protein content of the filtrate (% of the DM)	35.3	32.1	34.0	26.9
Protein content of the cake meal (% of the DM)	30.1	28.2	27.3	30.8
Protein yield in the filtrate (%)	**35.4**	**26.6**	**35.9**	**27.7**
Protein yield in the clarified filtrate (%)	**25.7**	**16.0**	**25.9**	**22.4**
Intensity (Amps)	22	10	10	12
Specific Mechanical Energy (Wh/kg)	60.8	68.0	67.9	63.5

The solid residue, obtained after centrifugation is called the "foot", while the supernatant is the "clarified filtrate".

The mass balance difference, for any trial, is less than 5 %, which is acceptable for such treatment.

The dry matter recovery in the filtrate was calculated from:

DM yield (%) = DM_F/DM_A x 100, where DM_A is the quantity of alfalfa dry matter introduced into the extruder during one trial, DM_F is the quantity of dry matter in the filtrate obtained from the trial.

The protein extraction yield in the filtrate was calculated from:

Protein yield (%) = P_F/P_A x 100, where P_A is the quantity of alfalfa proteins introduced into the extruder during one trial and P_F is the quantity of proteins in the separated filtrate for the same trial.

The ash recovery in the filtrate was calculated from:

Ash recovery (%) = A_F/A_A x 100, where A_A is the ash content of the alfalfa introduced into the extruder during one trial and A_F is the ash content of the filtrate for the same trial.

Alfalfa extrusion without water addition (trial 18) leads to the production of a partially dehydrated cake meal, since the dry matter content increases from 20 % (initial value in the plant) to 27 %. This treatment allows recovery of 26 % of the alfalfa proteins. Nevertheless, part of these are still linked to cell walls, and/or are insoluble, since the foot content is about 20 %, and the protein extraction yield decreases to 16 % after centrifugation.

When the L/S ratio is increased to 7.6 % (Trial 19), alfalfa dehydration remains similar to the previous results (about 26-27 %), although with a better dry matter recovery yield, increasing from 14 % to 22 %, after centrifugation. The main difference with trial 18 is the lower foot content recovered in the filtrate, decreasing from 21 % to 14 %. The protein loss during centrifugation is thus lower in the trial 19 filtrate. This difference could be attributed to higher filling of the screw spaces, and a higher residence time in the extruder. Thus, an experiment has been carried out at a similar L/S ratio, but with a higher alfalfa flow rate (trial 17). The results obtained are very close to the previous one (trial 19), showing the influence of extruder filling.

Water addition does not lead to a decrease in dehydration efficiency, because water induces washing of the plant cellulose fibers, and removes the extracted molecules (proteins, hemicelluloses, pectins), which are crucial because of their lubricating properties, leading to the reduction of the efficiency of the plant squeezing step.

The analysis of the dry matter and the moisture content repartition along the extruder axis is shown on Figure 9.

The moisture content of the plant matter at the entry of the extruder is higher than that of the frozen alfalfa (about 85 % of moisture in section A, 80.1 % of moisture content when frozen). This can be explained by the fact that the alfalfa is thawed at + 4 °C for a day, and can absorb water, thus increasing its moisture content. The first decrease in moisture in section B shows that the alfalfa is pressed by the monolobe paddle-screws. Their radial compression and shearing actions lead to accumulation of the matter, allowing a first pressing. However, the matter reabsorbs this liquid and also the water added in section D, as shown by the progressive increase in its moisture content, back up to its initial value, due to an efficient mixing zone (BB in module D). The matter then accumulates after the CF1TC screw, creating the dynamic plug required for pressing and liquid/solid separation, which explains the second decrease in solid moisture content in section G. The matter composing the dynamic plug is wetter for a high L/S ratio, even if the cake meal DM contents are similar, showing that a drier dynamic plug is not more efficient for L/S separation.

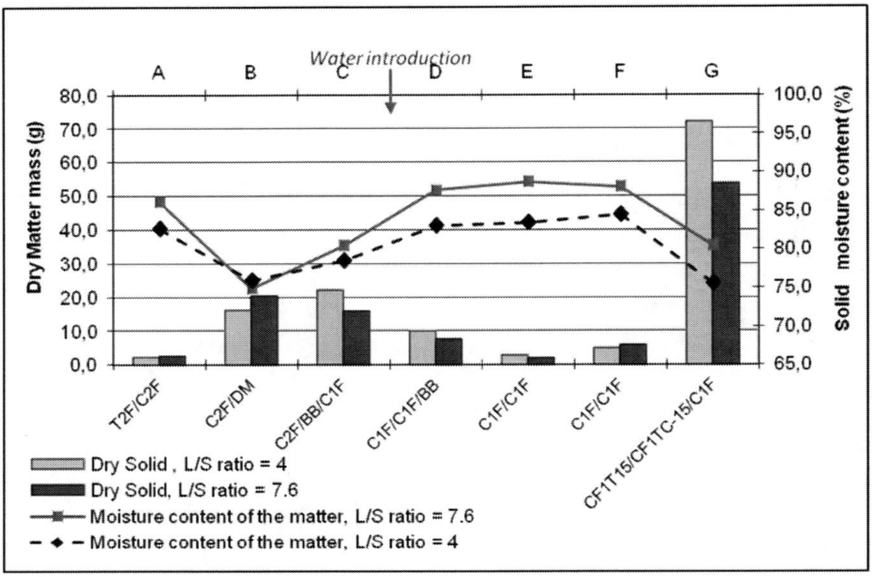

Total solid mass collected = 582.6 g (129.9 g of dry solid) for L/S ratio = 4; total solid mass collected = 585.3 g (108.1 g of dry solid) for L/S ratio = 7.6

Figure 9. Repartition of the matter and its dry matter content along the axis of the twin-screw extruder, for trial 18 (L/S ratio =4.0) and trial 19 (L/S ratio = 7.6), using screw profile 1.

Profile 2 allows removal of 62 % of the water of the alfalfa. Nevertheless, only 30 % dry cake meal content is achieved. As alfalfa seems to reabsorb the water expressed and added, a new screw arrangement, profile 3, has been investigated, with a shorter distance between the pressing zone and the sieve.

4.2. Screw Profile 3

The results obtained for the third screw profile with restrictive screw elements placed closer to the filter are shown in Table 5.

Table 5. Experimental results for protein extraction and alfalfa dehydration by CLEXTRAL BC 45 twin-screw extruder: Screw Profiles 3

Trial Number	**21**	**22**	**23**
Screw Profile	3	3	3
Liquid/solid Ratio	7.4	4.0	4.0
Alfalfa flow rate Q_A (kg/h)	12.3	15.3	32.9
Water flow rate Q_W (kg/h)	8.4	0.0	0.0
Filtrate flow rate Q_F (kg/h)	16.5	9.9	20.7
Cake meal flow rate Q_{CM} (kg/h)	4.5	5.7	11.2
Foot content of the filtrate (%wt)	14.4	19.7	18.9
Dry Matter content of the clarified filtrate (%)	3.5	5.5	7.8
Dry Matter content of the cake meal (%)	32.0	34.2	36.4
DM recovery in the clarified filtrate (%)	**20.6**	**15.8**	**20.6**
Protein content of the filtrate (% of the DM)	30.8	27.4	26.9
Protein content of the cake meal (% of the DM)	28.6	29.7	28.9
Protein yield in the filtrate (%)	**42.6**	**28.4**	**33.6**
Protein yield in the clarified filtrate (%)	**22.3**	**15.2**	**19.5**
Intensity (Amps)	10	12	17
Specific Mechanical Energy (Wh/kg)	65.0	62.9	41.4

The protein extraction yield in the filtrate is calculated using the total amount of proteins introduced into the extruder with the alfalfa.
DM: Dry Matter

Screw profile 3 allows removal of 70 % of the water introduced by alfalfa (trial 22, without water addition), whereas under similar conditions with profile 2 (trial 18) the value was only 50 %. The second profile is thus more efficient at dehydration, but not for protein recovery. Indeed, the protein recovery yield after centrifugation is only 15 % wt. More proteins are extracted with profile 3, but much of these are lost during centrifugation. As for profile 2, water addition increases protein recovery, without hindering dehydration efficiency, since the dry weight of the cake meal remains above 32 %.

For trial 23, extrusion has been undertaken without water addition, but at a high feed flow rate. The screw space is filled much more, increasing residence time of the matter, allowing better mechanical energy transmission. This test gave the best results for alfalfa dehydration, with removal of 72 %

of the water introduced with the plant (water content of alfalfa), leading to a 36 % dry matter cake meal. The protein extraction yield is 20 %, lower than that obtained in trial 18, although with a very low specific mechanical energy (41 $W.h.kg^{-1}$)

The protein extraction yields in the green juices are better with increasing addition of water into the extruder (42.6 % for trial 21, 28.4 for trial 22). Water introduction is the most significant parameter for protein recovery, due to its washing action solubilizing the proteins. Compared with the screw profile 2, it can be seen that a larger part of the proteins are lost during centrifugation, meaning that profile 3 leads to better extraction of non soluble proteins, removed from the extract during centrifugation.

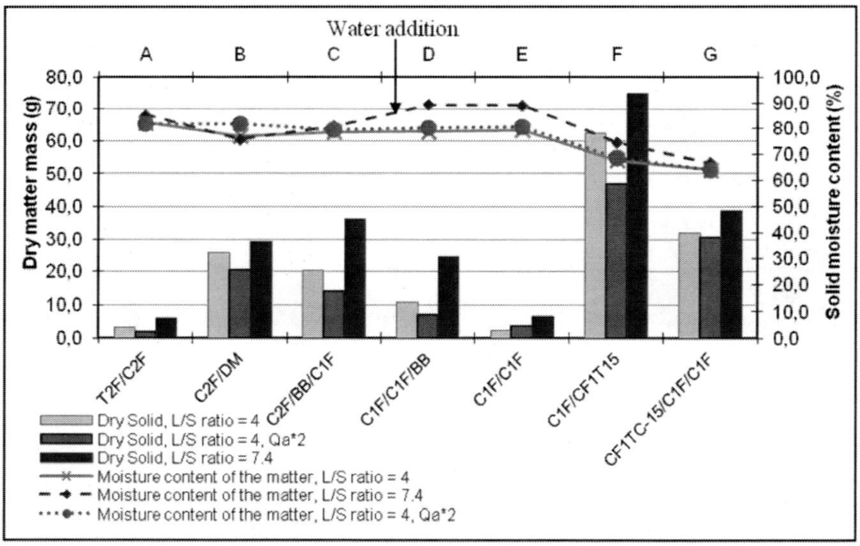

Total dry matter collected: 157.1 g for L/S ratio = 4; 124.8 g for L/S ratio = 7.4; 216.1 g for L/S ratio = 4 and alfalfa rate * 2.

Figure 10. Repartition of the matter and its dry matter content along the axis of the twin-screw extruder, for trial 21 (L/S ratio = 7.4), trial 22 (L/S ratio =4.0) and trial 23 (L/S ratio = 4.0, alfalfa introduction rate * 2), using screw profile 3.

Figure 10 shows the evolution of the dry matter and the humidity repartition along the extruder axis. At the highest L/S ratio (7.4), the trend of the evolution of moisture content is similar to the results obtained with the profile 2.

However, the decrease in the moisture content covers two and a half modules with profile 3, showing that the compression zone is longer (dry matter accumulated in the last two modules), and meaning that the mixing zone is correspondingly shorter. When no water is added in the extruder, this mixing zone is too small and the residence time too short to let the plant matter reabsorb the released water.

This water reabsorption phenomenon by the fibers seems to be stronger when plant destructuration is higher, which could be linked to an enhancement of the contact surface between plant material and water. However, this water is easily recovered if pressing is efficient, and if fiber leaching is sufficient. The mixing zone is a key parameter, because as the water is absorbed by the fibers, the proteins can be solubilized and then extracted when the plant is pressed. The quantity of water required to solubilize and extract the proteins depends on the alfalfa protein content. Evon et al., 2009 showed that the solid phase residence time is almost six times shorter than the water residence time in the mixing zone. Thus, a screw profile with more restrictive screw elements may increase the intensity of the mixing, generate more heat and thus increase residence time of the solid, allowing water absorption.

4.3. Screw Profile 4

Profile 4 has been designed with the same screws as in profile 3, but with two squeezing zones, associated with two filtering zones. This profile therefore allows recovery of a first extract, and substitutes this protein-rich water with fresh, added water. This profile can be compared to a two-stage extractor, and is used to determine whether protein solubilization in the water could be a limiting factor.

No trials without any external water addition were carried out, thus the aim of testing this profile was the second pressing step after water addition. The only difference between trials 24 and 25 (Table 6) is the increase in the L/S ratio, from 6.8 (trial 24) to 10.4 (trial 25). For trial 26, a L/S ratio of 10.7 is achieved, thanks to the double water addition.

The results obtained during trial 24 indicate that the dehydration efficiency remains high, similar to that obtained with profile 3. The dry matter recoveries are the same for each step, taking into account the initial content. But when the dry matter recovery yield is calculated from the second extraction only, it can be seen that this second step is more efficient than the first. The second stage allows the same quantity of proteins to be extracted as the first one (on matter already poorer in proteins), which indicates that protein recovery is hindered by their solubilization in water. However, the mechanical action is more intense with profile 4, leading to higher foot contents (higher plant destructuration) and thus to higher specific mechanical energy consumption.

Table 6. Experimental results for protein extraction and alfalfa dehydration by CLEXTRAL BC 45 twin-screw extruder: Screw Profile 4

Trial Number	**24**	**25**	**26**
Screw Profile	4	4	4
Liquid/solid Ratio	6.8	10.4	10.7
Alfalfa flow rate Q_A (kg/h)	15.1	14.2	14.3
Water 1 flow rate Q_{W1} (kg/h)	0.0	0.0	1.7
Water 2 flow rate Q_{W2} (kg/h)	8.6	18.3	17.5
Filtrate 1 flow rate Q_{F1} (kg/h)	6.7	5.8	7.6
Filtrate 2 flow rate Q_{F2}(kg/h)	12.3	22.4	21.6
Cake meal flow rate Q_{CM} (kg/h)	4.9	4.5	4.2
Foot content of the filtrate 1 (%)	18.2	17.8	12.5
Foot content of the filtrate 2 (%)	11.8	10.4	10.8
Dry Matter content of the cake meal (%)	32.8	34.2	36.4
DM recovery in the clarified filtrate 1 (%)	**14.5**	**12.8**	**14.8**
DM recovery in the clarified filtrate 2 (%)	**13.0**	**17.9**	**12.4**
Protein content of the filtrate 1 (% of the DM)	28.6	29.0	25.2
Protein content of the filtrate 2 (% of the DM)	28.4	25.2	32.7
Protein content of the cake meal (% of the DM)	28.9	26.5	26.6
Protein yield in the filtrate 1 (%)	**25.5**	**21.5**	**21.9**
Protein yield in the filtrate 2 (%)	**25.2**	**29.3**	**25.7**
Protein yield in the clarified filtrate 1 (%)	**14.6**	**13.1**	**13.1**
Protein yield in the clarified filtrate 2 (%)	**13.0**	**15.8**	**14.3**
Protein extraction yield in the clarified filtrates (%)	**27.6**	**28.9**	**27.3**
Intensity (Amps)	13	12	13
Specific Mechanical Energy (Wh/kg)	68.8	67.8	73.0

The protein extraction yield in filtrate 2 is calculated using the initial amount of proteins introduced into the extruder with the alfalfa.

Increasing the L/S ratio does not allow any significant improvement in matter dehydration (DM of the solid residue = 32.8 % for trial 24, 34.2 % for trial 25 and 36.4 % for trial 26). If water is only added through the first inlet, dry matter recovery in the first filtrate is better, but decreases in the second filtrate. Concerning protein extraction, increasing the water flow rate does not enhance protein recovery in the first filtrate (25.5 % of the proteins recovered in the first filtrate for trial 24, only 21.5 % and 21.9 % for trials 25 and 26, respectively).

Compared to the results obtained for the other screw profiles, the water flow rate in the second inlet, or the residence time in the liquid/solid extraction zone, seem to be insufficient here to ensure good washing of the fibers. Two successive filtrations allow the extraction of about half of the proteins from the alfalfa in the raw filtrate, with the same proportion of proteins extracted in each filtrate.

After centrifugation, the protein extraction yield decreases by 50 %, in each case. When the amount of water added is greater (trial 25, L/S ratio = 10.4), protein extraction is more efficient in the second stage. However, the mechanical action seems to be more intense, leading to the extraction of a larger amount of proteins that remain in the precipitate after centrifugation.

When, for the same L/S ratio, water is introduced in two steps (trial 26), no significant difference in the results occurs, except a slightly lower extraction yield in the second filtrate. The water content of the alfalfa seems to be sufficient to ensure protein extraction during the first pressing. The water added through the second inlet is useful for washing and solubilizing the proteins, which are more accessible due to the action of monolobe and bilobe paddle-screws.

Figure 11 shows the repartition of the matter along the axis, inside the extruder. After the first pressing, in module D, the matter is unrestricted in module E, and is then compressed in module F. With profile 4, the absorption area is longer than for the two other screw profiles, favoring proteins solubilization and thus their extraction. The same tendency can be observed Figure 11 (B) for the humidity curve for trial 26 (two additions of water), and also for the solid dry matter repartition. This shows that an extra water inlet, before the first filtration step, has no influence either on the mechanical action inside the extruder, or on the filling, or on the quality of the fractions.

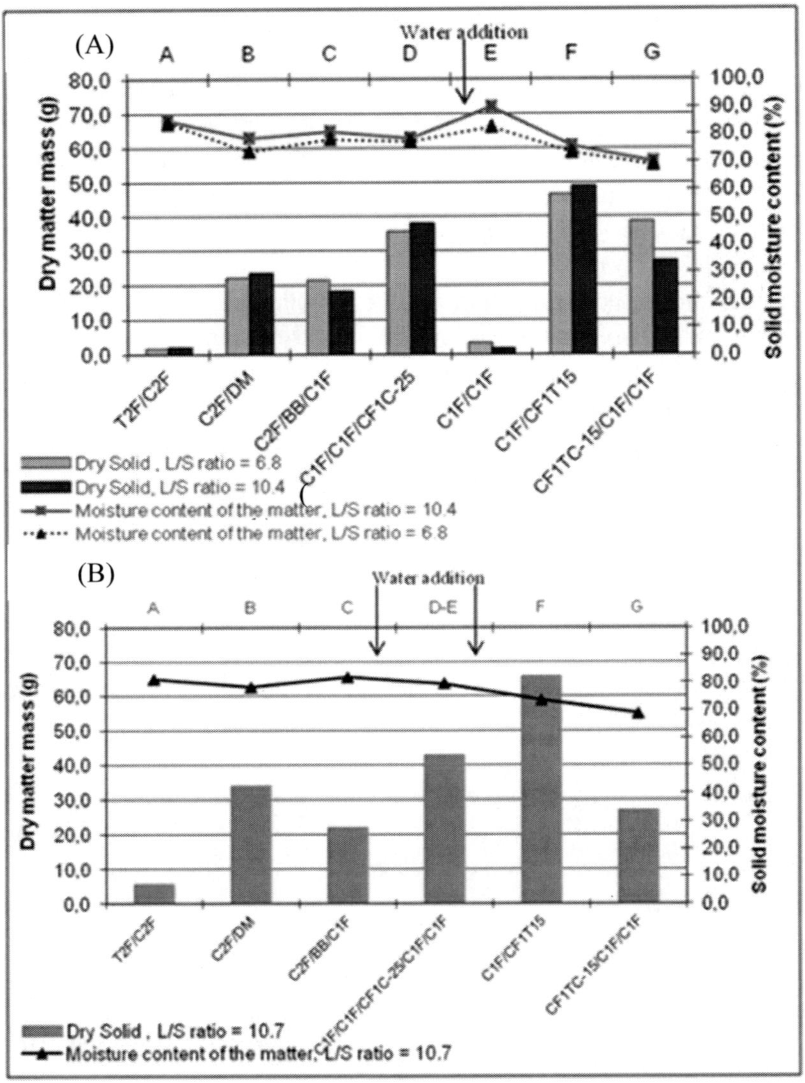

Figure 11. Repartition of the matter and its dry matter content along the axis of the twin-screw extruder, using screw profile 4. A) One water addition inlet: trial 24 (L/S ratio = 6.8, total dry matter collected: 169.2 g) and trial 25 (L/S ratio = 10.4, total dry matter collected: 159.4 g); B) Two water addition inlets: trial 26 (L/S ratio = 10.7), total dry matter collected: 197.4 g.

4.4. Comparison of the Results: Influence of Screw Profile on Alfalfa Dehydration and Protein Extraction

To compare the results obtained under the different twin-screw extruder configurations, statistical analyses were carried out. They consisted of one-way analysis of variance, for each tested parameter, with $\alpha = 5$ %. Analyses were performed with the Microsoft Office Excel 2007 software.

All the results obtained, for the three screw profiles, were compared, in order to determine if there is a significant influence of the screw profile configuration on several parameters (solid fibrous residue dry matter, protein extraction yield in the filtrate, protein outlet flow rate).

For the profiles evaluated, screw profiles 3 and 4 give the best alfalfa dehydration (Figure 12 (A)), with a residual water content of 55 % (compared to the initial alfalfa water content of 80 %). However, the results show that all screw profiles tested allow only partial dehydration of alfalfa, showing that extrusion alone cannot be used for alfalfa dehydration.

Extrusion appears to be an interesting technique for protein extraction (Figure 12 (B)). However, even if protein extraction yields in the filtrates are high, after centrifugation they are only about 20 to 30 %, and they do not fluctuate significantly from one trial to another. The highest extraction yield in the centrifuged extract is obtained for trial 24 with profile 4 (32 %), very close to the value obtained for trial 25, still with profile 4. Trial 20, although giving a lower protein extraction yield in the raw filtrate, is the trial with the lowest loss of proteins during centrifugation. However, the protein yield is calculated from the alfalfa total protein content, whereas only 50 % wt are soluble proteins (Douillard and De Mathan, 1994). Thus, the extraction yield of soluble proteins is about 60 %, with a very short residence time (less than 3 min).

On Figure 12 (C), it can be seen that profile 4, even if it is the most favorable screw configuration for protein extraction in the filtrate, is the worst for energy consumption. The differences between the three screw profiles for each parameter are shown in Table 7.

A significant difference was determined at a confidence level of 95 % between the influence of each profile on the alfalfa dehydration (dry matter content of the solid fibrous residue, P-value = 0.0031), and on the protein extraction yield in the raw filtrate (P-value = 0.0058). A significant difference was determined at a confidence level of 92.5 % between the profiles for the protein extraction yield in the clarified juice (P-value = 0.0510).

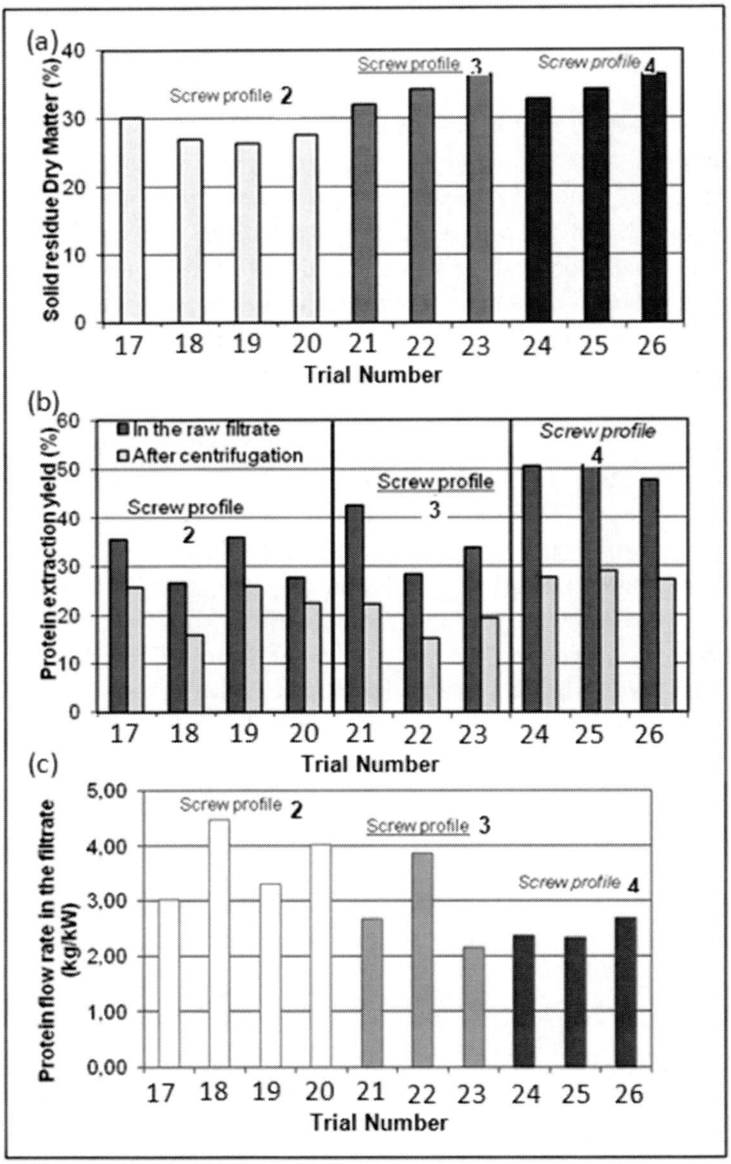

Figure 12. Influence of screw profile on: (a) Solid residue dry matter, (b) Protein extraction yield, (c) Protein flow rate in the filtrate.

Table 7. ANOVA to determine the effect of the screw profile (profile 2, profile 3, profile 4) on: solid residue dry matter, protein extraction yield in the filtrate, protein flow rate in the filtrate, protein content of the filtrate, percentage of proteins lost during centrifugation, protein extraction yield in the centrifuged juice

	p-Value	Significant influence of the screw profile (level of 0.05)	Significant influence of the screw profile (level of 0.075)
Solid residue dry matter (%)	0.0031	**Yes**	Yes
Protein extraction yield in the filtrate (%)	0.0058	**Yes**	Yes
Protein flow rate in the filtrate (kg/h)	0.6312	No	No
Protein content of the filtrate (%/MS)	0.7288	No	No
% loss of protein during centrifugation	0.0111	No	No
Protein extraction yield in the centrifuged juice (%)	0.0510	No	**Yes**

From these results and Figure 12, it is possible to conclude that the best screw profiles tested for alfalfa dehydration are profiles 3 and 4, for the second alfalfa batch. Moreover, profile 4 is also the best screw configuration for protein extraction in the raw filtrate. Considering the protein extraction yield in the centrifuged juice, the influence of the centrifugation step is open to question, because there is no significant difference between the profiles at a confidence level of 95 %. A finer filtration step, instead of centrifugation, could be more adapted for the elimination of the solid particles in the extract, without losing so much protein.

There is no significant difference between the profiles on protein content in the filtrate, or on the protein flow rate. This is due to the variability of the results using the same profile across the trials, showing that other parameters, such as alfalfa flow rate or L/S ratios, have as much influence here. A study of the local residence times in each zone of the extruder, for the liquid and for the solid phase, could provide more information in order to optimize the three zones in the twin-screw extruder (the mixing zone, the liquid/solid extraction zone and the liquid/solid separation zone), by slightly modifying the screw profile. The influence of alfalfa quality should also be taken into account, and the screw profile should be adapted accordingly, in order to achieve the best protein extraction yield and the best dehydration.

5. Twin-Screw Extrusion: A Step in the Alfalfa Biorefinery

As we have demonstrated in this work, twin-screw extrusion is an efficient tool to extract proteins from alfalfa whole plant in a filtrate that can be further fractionated and valorized. The dehydration of the plant is not sufficient with this technique, but twin-screw extrusion can be a first step for this. The method has to be integrated into a complete fractionation process of alfalfa, as in the example described below.

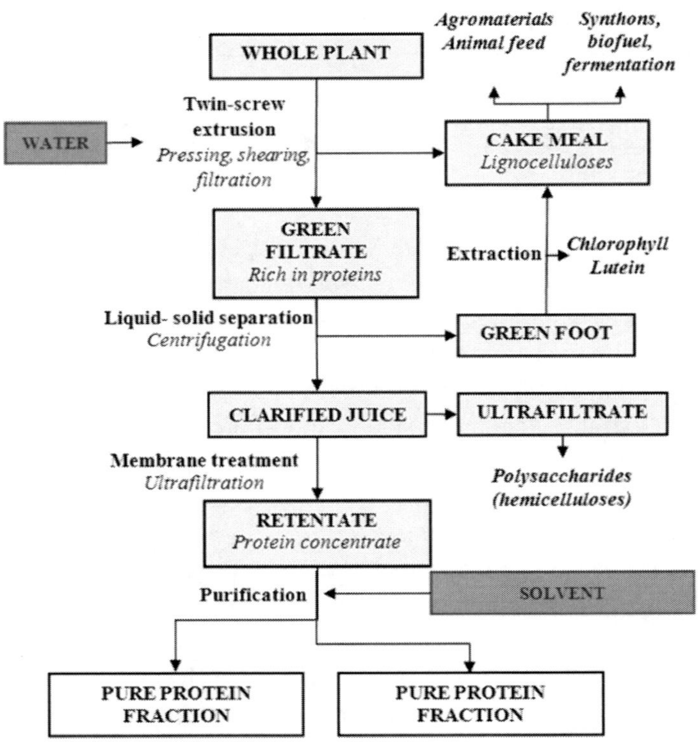

Figure 13. An example of a fractionation process for alfalfa.

Some steps of this fractionation process have been studied (data not shown). After twin-screw extrusion, centrifugation of the filtrate is necessary in order to remove solid particles, which could hinder the efficiency of further purification steps, e.g., ultrafiltration or chromatography. After centrifugation, a green, viscous foot is obtained, which could be dehydrated and used for animal feed as cake meal, because it still contains the proteins and fibers from the green extract. However, its deep-green color tends to show that this foot is nevertheless rich in high added value products, such as chlorophyll and lutein (Alves-Rodrigues and Shao, 2004).

Chlorophyll is industrially extracted from green vegetables, such as spinach. Production of this pigment could be carried out during the present fractionation process, from the green foot. To assess this hypothesis, a chlorophyll extraction process from the green foot and also from the green extract, was developed. Alfalfa was treated in acetone, chlorophyll was quantified using spectrophotometry, and calibration curves from synthetic solutions of chlorophyll a and chlorophyll b (Garcia et al., 1995; Porra et al., 1989).

Results are shown in Table 8:

Table 8. Determination of chlorophyll a and chlorophyll b content in alfalfa fractions

Sample	Chlorophyll a ($mg.g^{-1}$)	Chlorophyll b ($mg.g^{-1}$)	Total chlorophyll ($mg.g^{-1}$)
Green extract	1.5 ± 0.3	2.7 ± 0.6	3.9± 0.8
Green foot	20.9 ± 0.8	35.5± 0.4	57.0 ± 1.3
Spinach[1]	*n.d.*	*n.d.*	1.25 – 2.23
Lettuce[1]	*n.d.*	*n.d.*	0.42 – 1.14
Chinese cabbage[1]	*n.d.*	*n.d.*	1.16 – 2.04

[1]Yin et al., 2007.

The green foot, obtained after centrifugation of the green extract, is ten times more concentrated in chlorophyll than the filtrate, showing that almost all of this pigment is recovered in the foot after centrifugation. Chlorophyll content is related to the quantity of membrane proteins with photosynthetic properties (Garcia et al., 1995), and these proteins, still linked to the plant particles, precipitate during centrifugation.

Green foot is thus an important source of chlorophyll and could be valorized, on an industrial scale, since its concentration is 25 times higher than the traditional sources. A study of this step should be made in order to determine the best way to valorize this green foot: dehydration then grinding for animal feed, solvent treatment for chlorophyll extraction, or a mix of these two steps, with chlorophyll extraction then solvent removal before dehydration. The solid residue could be further dehydrated, for usual valorization, or directly fermented for biofuel or synthon production.

CONCLUSION

Twin-screw extrusion could be an efficient tool to extract proteins from vegetable matter, while washing cellulose fibers, prior to further treatment. In this study on alfalfa, more than 50 % of the global proteins can be extracted in the liquid extract after extrusion. In addition, the fibrous residue could be further valorized for the usual animal feed applications, although with the necessity for another dehydration step (the extruder alone is insufficient), or by fermentation for biofuel production.

Twin-screw extruder design is a crucially significant parameter, and the influence of the screw profile has been demonstrated. When mechanical action on plant tissue is stronger, the protein recovery is better, although it leads to higher liquid retention. The configuration of the screw profile can be optimized, in order to improve this mechanical action on the matter, to achieve better protein extraction yield and lower residual water content. However, the optimization remains difficult, as we have shown that the variability of the alfalfa chemical composition influences the results (differences in dehydration efficiency and protein extraction yields with the same screw profile, for two different alfalfa batches). Nevertheless, best results were obtained working with young alfalfa, with high water content and screw profile 4, demonstrating the interest of double filtration.

A complete fractionation process has been proposed, in order to show that the twin-screw extrusion of alfalfa could be a first step in its treatment. Each of the co-products, such as the cake meal (animal feed), the green filtrate (purification of protein fractions, that could have high added value), or the green foot (chlorophyll), need however to be valorized, to match the concept of biorefinery.

ACKNOWLEDGMENTS

This research received financial support from the CP2D program (Chimie et Procédés pour le développement durable: Chemistry and Processes for sustainable development), of the French National Research Agency (ANR 07 CP2D 16-03).

REFERENCES

Alves-Rodrigues, A.Shao, A. (2004). The science behind lutein. *Toxicology Letters*, 150, 57-83.

Amalia Kartika, I., Pontalier, P.Y., Rigal, L. (2005). Extraction of sunflower oil by twin screw extruder: screw configuration and operating condition effects. *Bioresour. Technol.*, 97, 2302-2310.

Andersen, M.Kiel, P. (2000). Integrated utilisation of green biomass in the green biorefinery. *Industrial Crops and Products*, 11, 129-137.

Barbeau, W.E.Kinsella, J.E. (1988). Ribulose bisphosphate carboxylase/oxygenase (rubisco) from green leaves-potential as a food protein. *Food Rev. Int.*, 4, 93-127.

Choudhury, G.S., Gogoi, B.K., Oswat, A.J. (1998). Twin screw extrusion of pink salmon muscle and rice flour blends: effects of kneading elements. *Journal of Aquatic Food Product Technology*, 7, 69-91.

Colas, D., Doumeng, C., Pontalier, P.Y., Rigal, L. (2013). Twin-screw extrusion technology, an original solution for the extraction of proteins from alfalfa (Medicago sativa). *Food and Bioproducts Processing*, 91, 175-182.

Douillard, R.De Mathan, O., (1994). Leaf protein for food use: potential of Rubisco. In: Hudson, B.J.F. (Eds.), Developing sources of food proteins, London, Chapman & Hall, 307-342 pp.

Evon, P., Vandenbossche, V., Pontalier, P.Y., Rigal, L. (2009). Aqueous extraction of residual oil from sunflower press cake using a twin-screw extruder: Feasibility study. *Ind. Crops Prod.*, 29, 455-465.

FAO (2002). Food energy - methods of analysis and conversion factors. FAO Food and Nutrition Paper, ISSN 0254-4725, 1-92.

Firdaous, L., Dhulster, P., Amiot, J., Gaudreau, A., Lecouturier, D., Kapel, R., Lutin, F., Vézina, L.-P., Bazinet, L. (2009). Concentration and selective separation of bioactive peptides from an alfalfa white protein hydrolysate by electrodialysis with ultrafiltration membranes. *J. Membrane Sci.*, 329, 60-67.

Garcia, J., Perez-Alba, L., Alvarez, C., Rocha, R., Ramos, M., Blas, C.d. (1995). Prediction of the nutritive value of lucerne hay in diets for growing rabbits. *Animal Feed Science and Technology*, 54, 33-44.

Gautam, A.Choudhury, G.S. (1999a). Screw configuration effect on residence time distribution and mixing in twin-screw extruder during extrusion of rice flour. *J. Food Process. Eng.*, 22, 263-285.

Gautam, A.Choudhury, G.S. (1999b). Screw configuration effect on starch breakdown during twin-screw extrusion of rice flour. *J. Food Process. Preserv.*, 23, 355-375.

Gogoi, B.K., Choudhury, G.S., Oswalt, A.J. (1996). Effects of location and spacing of reverse screw and kneading element combination during twin-screw extrusion of starchy and proteinaceous blends. *Food Res. Int.*, 29, 505-512.

Kamm, B.Kamm, M. (2004a). Biorefinery - Systems. *Chemical and Biochemical Engineering Quarterly*, 18, 1-6.

Kamm, B.Kamm, M. (2004b). Principles of biorefineries. *Applied Microbiology and Biotechnology*, 64, 137-145.

Lamsal, B.P., Koegel, R.G., Gunasekaran, S. (2007). Some physicochemical and functional properties of alfalfa soluble leaf proteins. *LWT*, 40, 1520-1526.

Maréchal, P., (2001). Analyse des principaux facteurs impliqués dans le fractionnement combiné de pailles et de sons de blé en extrudeur bi-vis : obtention d'agromatériaux, Thèse de Doctorat, INP Toulouse.

Mathieu, D.Phan-Than-Luu, R., (1992). New Efficient Methodology for Research using Optimal Design (NEMROD) Software. *Lprai*, Centre St Gerome, University of Aix-Marseille.

N'Diaye, S., Rigal, L., Goyette, C., Vidal, P. (1996). Extraction of hemicelluloses from poplar using twin-screw reactor: influence of the main factors. Developments in Thermomechanical Biomass Conversion. *IEA Bioenergy,* 1, 756-764.

Novalin, S.Zweckmair, T. (2009). Renewable resources - green biorefinery: separation of valuable substances from fluid-fractions by means of membrane technology. *Biofuels Bioproducts & Biorefining*, 3, 20-27.

Porra, R.J., Thompson, W.A., Kriedemann, P.E. (1989). Determination of accurate extinction coefficients and simultaneaous equations for assaying chlorophylles a and b extracted with different solvents: verification of the concentration of chlorophyll standards by atomic absorption spectroscopy. *Biochimica et Biophysica Acta*, 975, 384-394.

Rigal, L. (2000). Twin-screw technology, a new tool for fractionation, and thermo-mechano-chemical conversion of the agroressources. in Proceedings of the first World Conference on Biomass for Energy and Industry, Sevilla, Spain, Science Publishers.

Roustel, S. (2000). Cuisson-extrusion des aliments. Techniques de l'Ingénieur, traité *Agroalimentaire*, F 3120, 1-8.

Silvestre, F., Rigal, L., Leyris, J., Gaset, A., (1999). Colle à l'eau à base d'extrait protéique végétal et procédé de préparation, EP 0997513, France.

Van Soest, P.J.Wine, R.H. (1968a). Determination of lignin and cellulose in acid detergent fiber with permanganate. *Journal of the American Oil Chemists' Society*, 51, 780-784.

Van Soest, P.J.Wine, R.H. (1968b). Determination of lignin and cellulose in acid detergent fiber with permanganate. *JAOCS*, 51, 780-784.

Volenec, J., Cunningham, S., Haagensen, D., Berg, W., Joern, B., Wiersma, D. (2002). Physiological genetics of alfalfa improvement: past failures, future prospects. *Field Crop. Res.*, 75, 97-110.

Wasserman, B.P., Wan, L., Chan, K., (1992). Molecular transformations of starch and protein during twin-screw extrusion processing of cornmeal. In: Ed., Dekker M. Inc. (Eds.), Food extrusion Science and Technology, New York, 325-333 pp.

Xie, Z., Huang, J., Xu, X., Jin, Z. (2008). Antioxydant activity of peptides isolated from alfalfa leaf protein hydrolysate. *Food Chem.*, 111, 370-376.

Yin, Y., Han, Y., Liu, J. (2007). A novel protecting method for visual green color in spinach puree treated by high intensity pulsed electric fields. *Journal of Food Engineering*, 79, 1256-1260.

Zeitoun, R., Pontalier, P.Y., Maréchal, P., Rigal, L. (2010). Twin-screw extrusion for hemicellulose recovery: Influence on extract purity and purification performance *Bioresource Tech.*, 101, 9348-9354.

In: Alfalfa
Editor: Marie L. Hirsh

ISBN: 978-1-62948-752-6

Chapter 2

HIGHLY EFFECTIVE RHIZOBIA-ALFALFA-GRASS INTERCROPPING SYSTEM

Rui Zong Jia[1,2,*], Wen Xin Chen[1,†] and Jing Hui Liu[1,‡]

[1]Key Laboratory of Biology and Genetic Resources of Tropical Crop, Ministry of Agriculture China, Institute of Tropical Bioscience and Biotechnology, Chinese Academy of Tropical Agricultural Sciences, Haikou, Hainan, China

[2]State Key Laboratory of Agrobiotechnology; Key laboratory of Agro-Microbial Resource and Application, Ministry of Agriculture, College of Biological Sciences, China Agricultural University, Beijing, China

[3]Agronomy College, Inner Mongolia Agricultural University, Hohhot, Inner Mongolia, China

ABSTRACT

Biological nitrogen fixation, especially symbiosis contributes the most of nitrogen nutrition source on earth. Symbiosis has great potential in agriculture application, to reduce chemical fertilizer, to save energy, and to promote sustainable agriculture.

* Corresponding author: Rui Zong Jia, No. 4 Xueyuan Road, Haikou, Hainan, China, 571101. Phone: +86 898 66890981; Fax: +86 898 66890978; E-mail: reasonspare@gmail.com.

† Wen Xin Chen, E-mail: wenxin@cau.edu.cn.

‡ Jing Hui Liu, E-mail: cauljh@sina.com.

We evaluated the rhizobia-alfalfa-grass (*Sinorhizobium meliloti – Medicago sativa –Bromus inermis*) as model to study the promotion of nitrogen fixation, remission of the ammonia depression, of which the new knowledge will be a solid theoretical principle for rhizobia-legume-grass system application for ancient-old agricultural practices. Detailed works as following: (A) proposed the screening procedures for high effective symbiotic rhizobia in matter of nodulation ability, nitrogen fixation efficiency and competition capability with native rhizobia. Four strains CCBAU 01290, CCBAU 01199, CCBAU 01314, CCBAU 83164 were confirmed in filed testing; (B) the significant variations in symbiotic efficiency between the 4 rhizobia and 3 host plant cultivars. CCBAU 01199 and CCBAU 01290 were first to know as broader-host arrange rhizobia with all 3 alfalfa cultivars. Inoculated rhizobia could also alter the soil born rhizobia nodulation ability and biodiversity; (C) rhizobia-alfalfa-grass interaction system demonstrated the significantly promoting the alfalfa nodules and two crops yield. The high effective strains CCBAU 01199 and CCBAU 01290 in intercropping system minimized the ammonia depression that high nitrogen (NH_4^+) concentration around the root will deactive the nitrogenase to reduce the N_2 to NH_4^+; (D) evaluation the biodiversity of soil-born microorganism community after inoculation in rhizobia-alfalfa-grass. Understood the biodiversity and dynamic change of microorganism community for 6 months after inoculation will not affected soil bacteria permanently.

INTRODUCTION

Alfalfa (*Medicago sativa* L.) is believed to be the only forage crop that was cultivated before the recorded history [1]. This plant was induced from ancient Kashmir into China around 126 AD by Mr. Qian Zhang (about 164 to 114 AD), an ancient Chinese diplomat who explored the Silk Road. Up to date, alfalfa has become one of the most important forage plants in the northern parts and many cultivars adapted to the local conditions have been obtained in China.

The alfalfa rhizobial isolates from Chinese soils had phenotypic or genetic characteristics different from those originated from other countries [2-3], demonstrating that the compatible of rhizobia have evolved in association with the alfalfa cultivars grown in different eco-regions in China, or they were indigenous to the Chinese soils.

Different from the cultivation in New World countries, such as the US, Canada, and Australia, the alfalfa seed inoculation with rhizobia is rare in China.

In relation to the enlarged cultivation of alfalfa and the emphasis of sustainable cultivation in China recently, inoculation of rhizobia has been accepted by more and more Chinese farmers. Considering the increased demand of inoculants and the estimation that nodule symbiosis is a result of interaction among the host legume, the rhizobia and the environmental factors [4], some studies about the high efficient strains have been performed in China [5-7]. A separate inoculation of 10 *Sinorhizobium meliloti* strains could increase 80-250% (with average of 216%), 16.30-51.0 (2.1%) and 20.41-66.70% (with average of 29.9%) of alfalfa dry weight in three different cultivars [6], and 10-13% or lower in another experiment with 18 strains against the cultivar Vector (Canada) [5].

Therefore, it was clear that each alfalfa cultivar could select a subpopulation of *S. meliloti* as its suitable symbiotic partner. In China, the alfalfa cultivated areas are divided into three ecological regions according to the winter hardiness of alfalfa cultivars grown there [8]. The first eco-region included the northern Xinjiang and the central and western zones of Inner Mongolia where the altitude is 1000-2000 m; annual average temperature is 5-9 °C; annual precipitation is 50-250 mm; the winter is very cold and the summer is very hot. The second eco-region refers to the southern Xinjiang, where is very dry and hot in the growing season, with dramatic change of temperature in day/night time, and strong solar radiation. The other eco-region is intermediate type including the other Chinese provinces in temperate region, where has warmer and semiarid climate. Alfalfa cultivars with different morphologies and adaptation features have been found in the three eco-regions.

Inner Mongolia has the second biggest meadow in China and alfalfa is a popular cultivated forage plant there. In addition, the evaluation of effectiveness and competition of the rhizobial inoculants associated with the cultivars grown there is rather insufficient. Part of objects in this study was screening the high effective and competitive rhizobia adapted to soils of Inner Mongolia with a rapid and effective protocol.

A limit of available land resources has meant that high intensity cropping systems, such as intercropping, have emerged as one of the important strategy to increase the agricultural production [9]. Intercropping, simultaneously growing two or more crops in the same field is a traditional agricultural practice in China and in many in other countries, such as India, Mexico, and some African countries. The intercropping of squash and legumes was recorded in the earliest Chinese agricultural book "Fan Sheng Zhi Shu" written by an ancient agricultural officer, Fan Sheng Zhi in the 1st century BC.

In another ancient Chinese book "Qi Min Yao Shu" written by Jia Si Xie in the 6th century, some combinations of intercropping was described, such as mulberry with mung bean or adzuki bean. A primary reason for using the legumes as the dominant component of intercropping, such as in soybean/ maize and pea/wheat in China, or common bean/maize in Mexico, is that the legumes form root nodules with the symbiotic nitrogen-fixing bacteria, collectively named rhizobia [10]. This root nodule symbiosis is a most efficient biological nitrogen fixing system, which contributes produces the N source for the rhizobia themselves and the host plants, as well as for other microbes and plants growing in the same field simultaneously or subsequently [11].

Although more demands are being placed on agriculture there is little chance of increasing the amount of land currently in agricultural production due to competition from housing and other infrastructure. Also, forage crops are in direct competition with food crops. Thus, intercropping and mixed cropping systems have played an important part in increasing output of forage crops. Alfalfa is one of the most important legumes used in agriculture. Like other legumes, its root nodules contain nitrogen-fixing bacteria (rhizobia), in this case, *Sinorhizobium/Ensifer meliloti*. These bacteria fix atmospheric nitrogen which is then available for their own use and also available to the host plant and other soil microorganisms. Comparing to chemical fertilizers, biological nitrogen fixation appears to have many advantages. It acts in a specific manner, improving the legume-rhizobium interaction and N-fixing activity [12].

Rhizobial inoculation has been shown to increase biomass production both in terms of quality and quantity and also to enhance soil N and organic matter [13]. Agricultural practice has shown that legume cultivation increases the occurrence of rhizobia in the soil, and also that the higher the number of rhizobia in the soil the greater the yield of the legume [14]. However, one aspect often overlooked is that of "Ammonia Suppression". Higher concentrations of the NH_4^+ suppressed the nitrogenase reduction activity, while lower NH_4^+ concentrations triggered nitrogenase activity. Use of chemical fertilizers may negatively affect soil rhizobia, decreasing the number of nodules and nitrogen fixation efficiency [15] and may also have an impact on the soil microbial community [16]. It is important to conduct more studies to evaluate and assess the dynamics of the microbial community after soil has been inoculated with rhizobia.

There are many methods which have been used to study microbial communities.

These include traditional culture-based methods, such as serology [17], fluorescent antibody technique [18], enzyme linked immunosorbent assay (ELISA) [19], cell lipopolysaccharide [20], total protein patterns by SDS-PAGE [21], small subunit RNA sequence [22], plasmid profiles [23], multi-locus enzyme electrophoresis [24], more sensitive DAN-based methods including RAPD (Random amplified polymorphism DNA) [25], AFLP (Amplified fragments length polymorphism) [26], BOXA1R-PCR, ERIC-PCR, rep-PCR [27], and low molecular weight RNA profiling [28]. However, none of these methods are ideal due to problems with background activity, inability to culture samples, low sample concentration, poor repeatability, and interference from exogenous genes. Therefore, more sensitive and distinctive methods, such as DGGE (denaturing gradient gel electrophoresis) [29], T-RFLP (Terminal restriction fragment length polymorphism) were developed to allow for assessment of microbial diversity within a community as well as comparative distribution across communities [30]. An excellent application of T-RFLP in the microbial community was reconstruction of the RDB (Ribosomal Database Project) [30], which provided more information about the biodiversity of communities, and allowed rapid and high throughput identification. In this study we screened for highly effective symbiotic alfalfa rhizobia based on their nodulation, nitrogen fixation, competition with indigenous soil rhizobia, and assessed biomass accumulation in a field trial. We have also studied the effect on biomass accumulation caused by different combinations of alfalfa cultivars/rhizobial isolates, and the influence those combinations have on native rhizobial diversity and nodulation capability. We found that intercropping with alfalfa and siberian wild rye altered the soil enzyme activity and also the composition of the soil community [31]. We evaluated the influence of alfalfa intercropping with smooth brome grass (*Bromus inermis*) with and without rhizobial inoculation. Hypothetically, intercropping could affect symbiosis as well as the soil microbial community, which can be estimated by monitoring the dynamics of the microbial biodiversity of the rhizobia-alfalfa- smooth brome agricultural system.

MATERIALS AND METHODS

Strains and Their Geographic Origins

Ninety-six rhizobial strains were obtained from the Culture Collection, Beijing Agricultural University, China (CCBAU) and they were stored in 40%

(w/v) glycerol at -80 °C. These strains were isolated from nodules of *Medicago and Meliloti* grown in the northern regions of China, including provinces of Inner Mongolia, Shaanxi, Shanxi, Qinghai, Jilin, Gansu, Hebei, and Xinjiang (Table 1).

The strains were firstly activated by striking in YMA medium (pH 7.0-7.2) supplied with 0.02% BTB (Bromothymol blue) as pH indicator [32] and incubated at 28 °C. The purity of each strain was checked by observation of colonial morphology and cellular morphology with Gram-staining. Production of acid or alkali was recorded from the color change of BTB.

Experimental Site and Alfalfa Cultivars

A field experiment was established in April 2005 at the Institute of Horticulture, Inner Mongolia Municipality (E117.7’, N40.8’), where the annual temperature was 4-6 °C and annual precipitation was 300 mm, with 120 frost-free days. The soil properties were: loam, organic content 2.3%, alkali-hydrolysable nitrogen (N) 1.4 mg/Kg, available phosphorus (P) 41mg/Kg, available potassium (K) 144 mg/Kg, and pH 7.6.

Cultivation records showed that during the last 25 years there was no alfalfa or any other legume crops grown and the ground was irrigated but no fertilization before planting. The indigenous alfalfa rhizobia in this site was about 100 colony-forming units (cfu) /g of dry soil (data not shown), estimated by the alfalfa plant trapping method and calculated by mostly probability number (MPN) method [32].

Alfalfa cultivars Haygrazer, Vector and Aohan were kindly provided by Dr. Zhaohai Zeng (College of Agronomy and Biotechnology, China Agricultural University). The average of seed germination rate for each cultivar is 90-95%.

Screening Step I (Effectiveness Assay in Vermiculite in Greenhouse)

Inoculation. In this analysis, seeds of *M. sativa* cv. Aohan were used because this cultivar is native to Inner Mongolia and is one of the main alfalfa cultivars grown in the temperate region of China. The seeds were surface sterilised following the method of [32] and were sown in vermiculite moisturized with low-N nutrient solution [32] in Leonard jars [33].

Table 1. Rhizobial strains used in this study

Rhizobial strain (CCBAU no.)[a]	Eco[b]	Host species[c]	Resource
Sinorhizobium meliloti			
03041	III	*Med. sativa*	Jia 2009
83065, **83066**, 83068, **83085**	II	*Med. sativa*	Yan et al. 2000
83164, **83168**, **83178**	I	*Med. sativa*	Yan et al. 2000
Unclassified strains			
01001, 01135, 01187, 01262	I	*Med. falcata*	Inner Mongolia
01006, 01034, 01132, 01192, 01264, **01290**, 01291, 01298, 01300, 01301, 01302, 01306, 01307, 01308, 01309, 01310, 01311, **01314**, 01315, 01316, 01317, 01325, 01326, 01327, 01349, 01352, 01353, 01354, 01355, 01356, 01360, 01415, 01416, 01417	I	*Med. sativa*	Inner Mongolia
01013	I	*Mel. alba*	Inner Mongolia
01019, 01028, 01059, **01199**, 01236	I	*Med. lupulina*	Inner Mongolia
01280	I	*Mel. officinalis*	Inner Mongolia
03009	III	*Med. lupulina*	Shanxi
03035	III	*Med. Falcate*	Shanxi
05015, **05188**	III	*Med. sativa*	Hebei
05057, **05180**, **05210**	III	*Med. lupulina*	Hebei
05120	III	*Med. falcata*	Hebei
13012	III	*Med. falcata*	Jilin
13038, 13124 (nod-)	III	*Med. sativa*	Jilin
71001	III	*Med. minima*	Shaanxi
71004	III	*Med. sativa*	Shaanxi
71005	III	*Med. lupulina*	Shaanxi
71121	III	*Med. polymorpha*	Shaanxi
75006	III	*Med. sativa*	Ningxia
75011	III	*Med. falcata*	Ningxia

Table 1. (Continued)

Rhizobial strain (CCBAU no.)[a]	Eco[b]	Host species[c]	Resource
Sinorhizobium meliloti			
75035	III	*Med. sativa*	Ningxia
81062	III	*Med. archiducis-nicolai*	Qinghai
81007, 81099	III	*Med. lupulina*	Qinghai
83027, 83052, **83062**, 83080, **83083**, **96071**	II	*Med. sativa*	Xinjiang
83030	II	*Med. falcata*	Xinjiang
83028, 83031, 83032	I	*Med. falcata*	Xinjiang
83033	I	*Med.* sp.	Xinjiang
83025, 83050, **83165**, 83166, 83167, **83177**, 83181, **83535**, 83536, **83537**,**96287**	I	*Med. sativa*	Xinjiang

[a]Effective strains are in boldface and ineffective strains are underlined.
[b]Eco-type of alfalfa region cross China according to the (Lu, 1998).
[c]Host species *Med* represented *Medicago*, *Mel* represented *Meliloti.*

In each jar, 2-3 seeds were put and germinated in dark at room temperature for 48 h. Then the germinated seeds were inoculated separately with the 96 test strains (10^8 cfu/jar), and the blank controls were inoculated with sterilised water. The test strains were incubated in TY broth [34] with shaking at 28 °C for 48-72 h and the concentration was controlled at OD 500nm = 0.5 (equal to 1 x 10^8 cfu/mL). The experiments were performed in triplicate under natural sunlight and temperature in greenhouse, 14-16 h in daytime at 24-28 °C and 8-10 h in night time at 18-22 °C. All the bottles were arranged in randomized blocks, disarranged in every two days to avoid the variance caused by illumination difference, and watered with sterilising deionised water when it is necessary.

Biomass production and acetylene reduction assay. After 45 days of inoculation, the dry weight of shoots (DWO), total number of nodules (TNN), and activity of nitrogenase (acetylene reducing activity, ARA) were assayed. The shoots were cut and dried at 65 °C to constant weight and weighted. The roots were washed to eliminate attached vermiculite, soaked up on paper towel and then shifted into 100 mL serum bottle sealed with silica gel stopper for acetylene reduction assay immediately as described [35]. The nitrogenase activity was measured as following: drawn out 10 ml air from the serum bottle with roots, injected into 10 ml pure acetylene, incubated at 28 °C for 1 h. Then 100 μL gas sample was taken from the serum bottle and was injected into Hewlett Packard GC system (HP 6890 series, Hewlett Packard, INC, US) for gas chromatograph analysis. The conditions of GS configured as follow: injector temperature: 120 °C, with capillary column of HP plot / Al_2O_3 15 m x 0.53 mm x 15 μm film thickness (US), ultra high purity N_2 as carrier, with column flow 1.5 ml/ min, oven temperature 70 °C, FID detector at temperature 250 °C, H_2 flow 40 ml/ min, air flow 300 mL/min, makeup (N_2) flow 28.5 mL/min. The amount of ethylene calculated with the formulas F1 and F2: F1 $V = K \times C \times$ peak height ratio (C_2H_4/C_2H_2) where V is the volume of ethylene, K is the reciprocal related coefficient of peak height ratio (C_2H_4/C_2H_2), C is the volume of acetylene injected; F2 $N = V \times 1/22.4 \times 273/(273 + t\ °C) \times P/760 \times$ 106 where N presents nanogram molecule C_2H_4 at the standard atmospheric pressure (760 mmHg); t is 25 °C. The unit of nitrogenase (U) was defined as nanogram molecule C_2H_4 per hour per plant (root).

Screening Step II (Competition Assay in Soil in Greenhouse)

Inoculation. The soil used in this study was sampled from Institute of Horticulture of Inner Mongolia (Hohhot, capital of Inner Mongolia, 669 km north of Beijing). In this assay, alfalfa seeds without surface sterilisation were sown in Leonard jars (2-3 seeds each bottle) half filled with natural soil. Five duplicates were included. After two days for germination, 1 x 10^8 cfu of rhizobial culture was inoculated to each jar. The germination, inoculation and naturalisation of greenhouse were the same as mentioned in screening step I, except that tap water was used instead of low-N nutrition and sterilising deionised water.

Estimation of nodule occupation by inoculated strain with BOX-PCR. Nodules of each plant grown in the soil were counted and collected in 1.5 mL microtubes on the 50th day after inoculation. Twenty randomly selected nodules from one plant in each duplicate were surface sterilised and were used in bacteria isolation using YMA medium and protocols described by [32]. The isolates obtained from nodules were then incubated in TY to logarithmic phase for total DNA extraction following the method of [36]. DNA samples were stored at -20 °C until use. BOX PCR was performed in 25 μL reaction mixture with each DNA sample obtained from the nodule isolates by using BOXA1R primer (5'-CTA CGG CAA GGC GAC GCT GAC G-3') and corresponding protocol [27].

Electrophoresis was carried in 2.5% agarose supplied with 0.5 μg/mL ethidium bromide. The PCR fingerprinting was photographed and saved as TIFF format with Alpha Innotech Multimage (Alpha Ease FC v4.1.0, Alpha Innotech Corporation) after the electrophoresis.

The Gelcomp II software (v4.50, Applied Maths) was used to define the BOX PCR fingerprinting type and to cluster the isolates according to the BOX PCR patterns with Dice coefficient and UPGMA (unweighted pair group method with arithmetic mean) method was used to construct the dendrogram. The inoculated strains were included to compare with the isolates. Isolates sharing 100% similarity were identified as the same BOX type (strain).

Screening Step III (Nodulation Variation among the Alfalfa Cultivars in Field Trail)

Inoculation strains CCBAU 01199, CCBAU 01290, CCBAU 01314, and CCBAU 83164 were selected as highly effective alfalfa rhizobia, while CCBAU 13012 was used as a low effective control [34]. Each inoculum was harvested from the flask, washed twice with sterile saline to remove any media residue, homogenized and the bacterial concentration was adjusted to 10^{12} cfu/mL, The solutions were stored at 4 ^{o}C for a maximum of one week prior to inoculation. Field application was carried out by diluting the inoculums to 10^{2}-10^{3} cfu / seed with tap water.

Field trial design and management A randomized design was used for the 54 blocks covering all the treatments: 3 alfalfa cultivars × 6 inoculations (5 rhizobial + blank control) × 3 replicates. The area of each block was 12 m^2 (2 m×6 m). Row planting was carried out as row spacing 20 cm, depth 3-5 cm, sowing rate 15 kg/km^2.

Seed bed-based spray method described in previous study [37] was employed for sowing and inoculating the seeds. Fifty days after inoculation (DAI), five randomly selected plants at three random sites in each block were carefully uprooted and the height and weight of each shoot was determined. The number of nodules on each plant was also counted and 50 randomly selected nodules from each treatment were used for rhizobia isolation using standard procedure and YMA medium [32]. Subsequently an estimation of nodule occupation was made.

Total plant nitrogen and phosphorus determination All the shoots from each site were dried at 60 ^{o}C for about 48 h to a constant weight. Fine grounded plant powder was used for nitrogen and phosphorus analysis.

Nitrogen content was determined by Kjeldahl method [38], and phosphorus was determined by vanadium-molybdenum colorimetric method [39]. An aliquot of 0.5 g from each sample was used for analysis and each analysis was performed in triplicate. For nitrogen analysis an OD of 420nm was used [38] and for phosphorus analysis an OD of 450nm was used [39].

Nodule occupation determined by BOX-PCR fingerprinting Fifty randomly selected nodules from each treatment were surface sterilized and crushed on a YMA slant as described [32] and the isolates were purified by repeatedly striking on the same medium in Petri dishes. Each isolate, including the five inoculums, was incubated in 5 mL of TY broth to logarithmic phase for total DNA extraction and BOX-PCR [34]. The PCR patterns were visualized by electrophoresis in 2.0% (w/v) agarose with 0.5×TBE buffer. Photographs were taken with an Alpha Innotech Multimage system (Alpha Ease FC v4.1.0, Alpha Innotech Corporation) and saved as a TIFF format. The Gelcomp II package (v4.50, Applied Maths, Inc, US) was used to define the BOX PCR fingerprinting type and to cluster the isolates. Under the optimization set at 5% and position tolerance at 5%, PCR bands were defined and used to calculate the DICE coefficients which were employed to construct a UPGMA (Unweighted Pair Group Method with Arithmetic Means) dendrogram. Isolates sharing DICE coefficient 99% were believed as the same strain. The occupation of nodules by each of the inoculums was estimated.

Intercropping System Evaluation and the Soil Microorganism Community (Greenhouse Study)

Inoculation and management Alfalfa specific rhizobia *Sinorhizobium/ Ensifer meliloti* CCBAU 01199 was selected as a highly effective strain. The host alfalfa cultivar Algonquin was introduced to Inner Mongolia from Canada. Smooth brome grass (*Bromus inermis* Leyss), a local grass cultivar was chosen as the intercropping partner.

The plastic pots modified from the Leonard jars were sterilized by immersing in hydrochloric acid (1%, w/v) overnight, then H_2O_2 (3%, v/v) solution for 30 min and finally rinsed 5 times with sterile water. The jars were filled with soil (per 10 kg) supplemented with 0.02 % K_2HPO_4, 0.02 ‰ H_2B O_3, 0.02 ‰ H_2MO_4 and stabilized in the greenhouse for up to one week before planting.

In each row, 3-5 alfalfa seeds or 4-6 smooth brome seeds were sown in each row, and row spacing was 15 cm.

In the monocropping, all rows were sown with the same plants; while for intercropping, alfalfa and smooth brome were sown next to each other. The depth of sowing was 2 cm. Two days after germination, about 1x 10^3 cfu of rhizobia were inoculated to each seedling, while the same volume of sterilized water was added to each the control seedling as control. Ten replicates were performed for each treatment.

Throughout the experiment the soil moisture capacity was kept at 50-65 % with city tap water except for the first irrigation before planting (80 % moisture capacity). Randomly rearranging the jars twice a week avoided variations in illumination.

Soils sampling and soil DNA extraction The soil samples were collected six times: prior to inoculation (background control), 1, 2, 3 and 6 months after inoculation (MAIs). Collection at 3 MAIs coincided with the first mow and 6 months after inoculation was the time of the second mow. Soil samples were collected from three different sites in each jar at a depth of 10-15 cm. The soils were mixed evenly and stored at -80 °C prior to DNA extraction.

Soil DNA extraction was modified from a method previously described previously [34]. Unless specifically mentioned all procedures were carried out in an ice bath. Briefly, 0.5 g of soil was mixed in a 5 mL-tube with 500 μL of celite suspension solution by vortexing for 1 min., following by adding After this 1200 μL PBS buffer (phosphate buffered saline, 137 mM NaCl, 2.7 mM KCl, 10mM Na_2HPO_4, 2.0 mM KH_2PO_4, pH 7.4), 200 μL TE buffer (10 mM Tris-HCl, 1 mM EDTA, pH 7.5), and 460 μL TNS buffer (200 mM Tris-HCl, 50 mM EGTA, 0.5% SDS, 250 mM NaCl, pH 8.0) were added. Samples were then sonicated (JY92 Ultrasonic Cell Crusher, Ningbo Scientz Biotechnology Co., Ltd) at 100 watt, 10 cycles of sonication for 2 sec with 3 sec intervals. After centrifugation for 10 min at 15000x g the supernatant was transferred to new tubes and the soil pellets were resuspended in 1200 μL PBS buffer. Sonication, as described above, was then repeated. Supernatants were pooled together and to the tubes were added and 1.5 mL 7.5 M NH_4AC, was added, mixed well and then the same volume of ice-cold isopropanol was added. DNA was precipitated at -20 °C for 4 h, and pelleted by centrifugation at 15000x g for 30 min. To remove salt residues, the pellet was washed with 500 μL 75% ethanol with centrifugation at 15000x g for 5 min, and the pellet finally dried by vacuum desiccation (Vacufuge vacuum concentrator, Eppendorf). DNA was re-suspended in 50 μL Milli-Q water with RNase A (10 ng/mL). Purity and integrity was verified by agarose gel electrophoresis (Figure S2) stained with ethidium bromide.

T-RFLP analysis Bacterial 16S rDNA was amplified from soil DNA using 5-terminal fluoresce FAM labeled primer Bac8f: 5'AGA GTT TGA TCC TGG CTC AG-3' and primer 1492r: 5'-GGT TAC CTT GTT ACG ACT T-3'. The PCR reaction consisted of 25 μL 2x PCR master mix, 10 p M forward and reverse primers, 1 μL (~100 ng) DNA template and Milli-Q water to a final volume of 50 μL. Amplification was performed using a PTC-200 (MJ, Bio-Rad), with denaturation at 95 °C 10 min, followed by 10 cycles at 94 °C 1 min, 55 °C 1 min with 0.3 °C decrease /cycle, and 72 °C 1.5 min, then 20 cycles at 94 °C 1 min, 50 °C 1 min, and 72 °C 1.5 min, and a final elongation at 72 °C for 10 min. Products were visualized during electrophoresis by staining with ethidium bromide (Figure S2). For soil 16S rDNA polymorphism analysis, DNA was first digested using MspI and HhaI (NEB) according to manufacturer's instructions. To the digested DNA, 0.1 volumes of 2M NaAc (filter sterilized) and 1 volume of anhydrous ethanol were added, vortexed and centrifuged at 13000x g for 20 min. The supernatant was discarded before addition of 80 μL 70% ethanol, and centrifugation at 13000x g for 20 min.

Again the supernatant was discarded and vacuum desiccation used to completely remove the ethanol. The DNA fragments were re-dissolved in deionized formamide for 30 min atin room temperature and mixed with internal reference ROX (Genscan -1000 ROX Size standard, Applied Biosystems). A final step was to denature the samples at 95 oC for 5 min and this was performed prior to the fluorescently labeled terminal restriction fragments (TFs) being analyzed on an ABI 3130 Genetic Analyzer (Applied Biosystem) in Genescan mode. Statistical analysis followed previous methods [30].

Data Analysis

All data were analyzed by SAS (for windows v8, the SAS Institute Inc., Cary, NC, US), and EXCEL 2002 (Microsoft corporation, 2001-2003). The effectiveness of strains was assessed with PROC ANOVA of SAS by Duncan's Multiple Range Test performed at significance level of P=0.01 for all statistical tests. The Principal Component Analysis of SAS was performed to evaluate the predominance of different rhizobia and alfalfa combinations. The Canonical Correlation Analysis was used to evaluate the relationship between the biomass contribution and nodulation and competition. Shoot height, dry weight, total plant nitrogen and phosphorus were considered as BIOMASS contributors (the Canonical variable: BIOMASS).

The total box-types of the nodule isolates and the nodule occupation of the inoculums in each treatment were considered as biological nitrogen fixation index (the Canonical variable: SYMBIOSIS). The significance level P=0.01 was used. All the nodulation and growth data were analyzed by SAS for windows v8 (The SAS Institute Inc., Cary, NC, US), and EXCEL 2002 (Microsoft corporation, 2001-2003). The variance in effectiveness of the strains was analyzed with PROC ANOVA of SAS, in which Duncan's Multiple Range Test was applied and alpha = 0.01 was preferred to provide a 99% confidence limit. The principal component analysis of SAS was used to evaluate affection of the different inoculants on the promotion the biomass accumulation and nodulation based on the alfalfa's dry weight of shoots (DWO), total number of nodules (TNN), and acetylene reducing activity (ARA). Each index had equal weights. Prin1 = 0.592792 DWO + 0.549094 TNN + 0.589147 ARA (cumulative proportion of eigenvalue difference = 84.72%); Prin2 = -0.356228 DWO + 0.834855 TNN – 0.419666 ARA (cumulative proportion of eigenvalue difference = 95.89%).

Nonlinear canonical correlation analysis of SPSS (SPSS 13.0 for Windows, SPSS Inc.) was applied to figure out the relations between the ecology type of strain's resource and the effective- ness for the host plant. All the strains were divided into three ecogroups based on their geographical origins: (I) Inner Mongolia and North Xinjiang; (II) South Xinjiang; and (III) other areas including Hebei, Shanxi, Shaanxi, Gansu, Qinghai and Jilin (Lu, 1998).

The DWO, TNN and ARA of different treatments were also divided into three grades with QUARTILE method (EXCEL), the high grade was more than 75% of total samples, which means high ability to promote the host plants, contrary, low grade was less than 25%, and the rest of samples (between 25 and 75%) was classed to intermediate grade.

RESULTS

Strains Characteristics

All the test strains formed colonies of 2-4 mm in diameter within 48 h on YMA, indicating that all of them are fast-growing rhizobial, despite of their different geological origins. All the strains are Gram-negative short rods.

All the strains produced acid that was evidenced by the yellow colour in YMA supplied with BTB [32].

Analysis of Screening Step I

Dry weight The dry weights of shoots of alfalfa grown in vermiculite inoculated with each of the 96 rhizobial strains are showed in Figure 1. In the statistical analysis, the dry weights of shoots inoculated with different strains were very different.

Most of strains showed plant growth promoting, and the host biomass was increased significantly than control ($P < 0.0001$) by inoculation of 24 strains, respectively isolated from Inner Mongolia, Hebei and Xinjiang.

The strains isolated from other areas such as Shanxi, Shaanxi, Gansu, Qinghai and Jilin showed lower effectiveness.

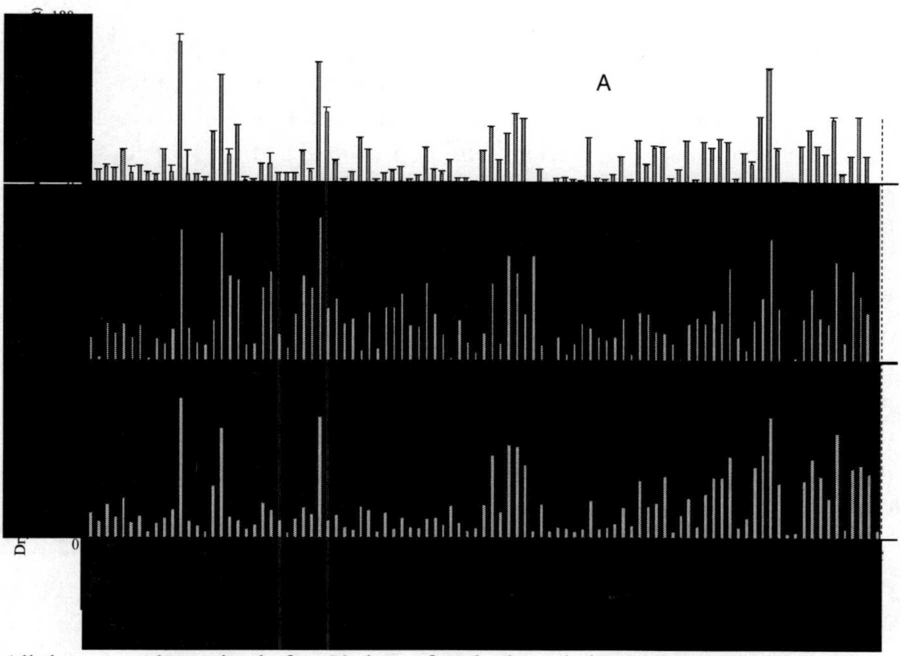

All data were determined after 50 days after the inoculation. All the strains were from the Culture Collection, Beijing Agricultural University, China (CCBAU).

Figure 1. Effectiveness of alfalfa rhizobial obtained in the nodulation tests in vermiculite. A: nitrogenase activity measured by acetylene reduction assay (nanogram molecular C_2H_4 per hour per plant). B: dry matter of overground (DWO) of alfalfa inoculated with different rhizobial strains. C: total nodule numbers of alfalfa inoculated with different rhizobial strains.

In addition, 11 strains respectively isolated from Jilin, Shaanxi, Shanxi, Inner Mongolia and Xinjiang were not effective on the alfalfa cultivar involved in this study (Figure 1 and Table 1). Some of them even caused yellow-green leaves, white freckle on surface of leaves, and rotten and brown roots in the inoculated plants.

Total number of nodules and acetylene reduction assay The results are summarised in Figure 1.

The total number of nodules and ARA showed a good correlation with each other and with the biomass production. Corresponding to the high dry weigh of shoots, the total number of nodules and ARA were also great. Additionally, the weaker ARA or smaller numbers of nodules were always related to the smaller biomass accumulation. Only one strain CCBAU 13124 from Jilin did not form any nodule on the host plants.

Nonlinear canonical correlation analysis The results of this analysis, presented in Figure 2, demonstrated that, over 99% confidence level, all the high and intermediate values of dry weights, nodule numbers, and acetylene reduction were closely associated with the rhizobial strains isolated from eco-region I, while the strains isolated from eco-regions II and III were closely related to the low values.

These relationships indicated that the rhizobial strains were more effective with alfalfa grown in the same eco-region.

Principal component analysis The principal component analysis of alfalfa biomass and nodulation showed that the test strains could be divided into three groups. Group 1 included 4 strains (CCBAU 01290, CCBAU 01199, CCBAU 01314, and CCBAU 83164) which were the most effective symbionts when all the data of biomass production, nodulation ability and ARA were considered. Group 2 included treatments with 23 isolates that had no significant difference with blank control. Group 3 covered the remaining strains that showed intermediate effectiveness (Data not show).

Analysis of Screen Step II (Nodule Occupation)

Nodule occupancy In this step, all the 4 strains in group 1, four randomly selected ineffective strains in group 2, and nine randomly selected intermediate effective strains in group 3 were used. To estimate the nodule occupancy, the BOX PCR fingerprinting of CCBAU 01290 is shown in Figure 3 as an example. The nodule occupancy of the test strains is presented in Figure 4.

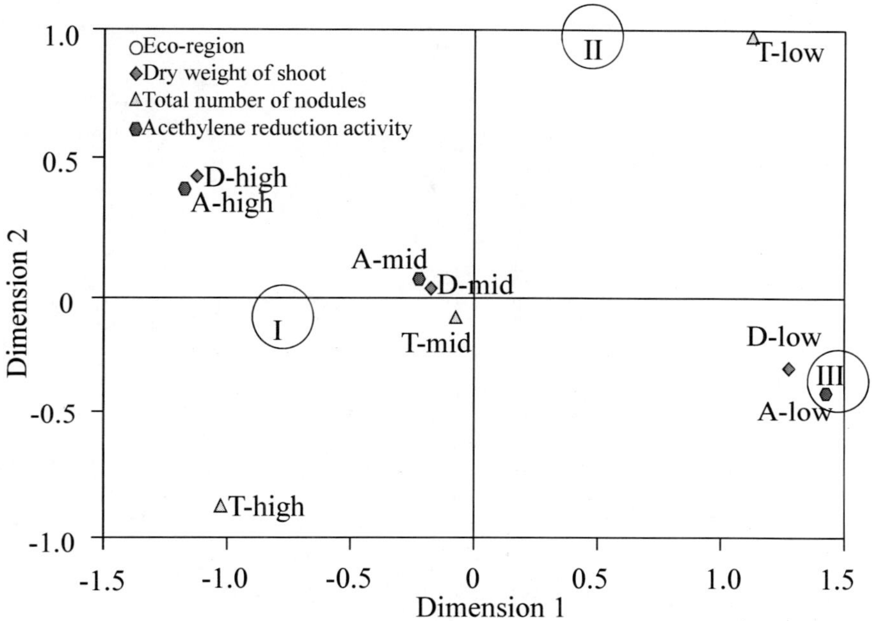

Figure 2. Nonlinear canonical correlation analysis among the eco-regions, the biomass accumulation, nodule numbers and nitrogenase activities. Confidence level was 99%. Optimal scaling level: ordinal. Three eco-regions (Inner Mongolia and North of Xijiang, South of Xinjiang, and other areas), three levels of increases (high = up to 75%, middle = 25-75% and low = below 25%), in dry weight accumulation (D), total numbers of nodules (T), and acetylene reduction activity (A) were used as levels for each factor. Dimension 1 and Dimension 2 may have no real meaning.

Based upon the statistical analysis, inoculation of the 4 strains in group 1 (CCBAU 01290, CCBAU 01199, CCBAU 01314, CCBAU 83164) and 2 strains in group 2 (CCBAU 83025 and CCBAU 83535) caused significant increase of nodule numbers of the alfalfa plants compared with the blank control.

However, only the 4 strains of group 1 and the group 2 strain CCBAU 83025 showed nodule occupancy more than 50%. These results demonstrated that the 4 most effective strains were also most competitive ones in the soil of Inner Mongolia.

Furthermore, CCBAU 83025 was no effective in vermiculite, but rather competitive in soil (similar to CCBAU 01199 and CCBAU 01314), demonstrating that effectiveness and competitiveness were not always associated in a single strain.

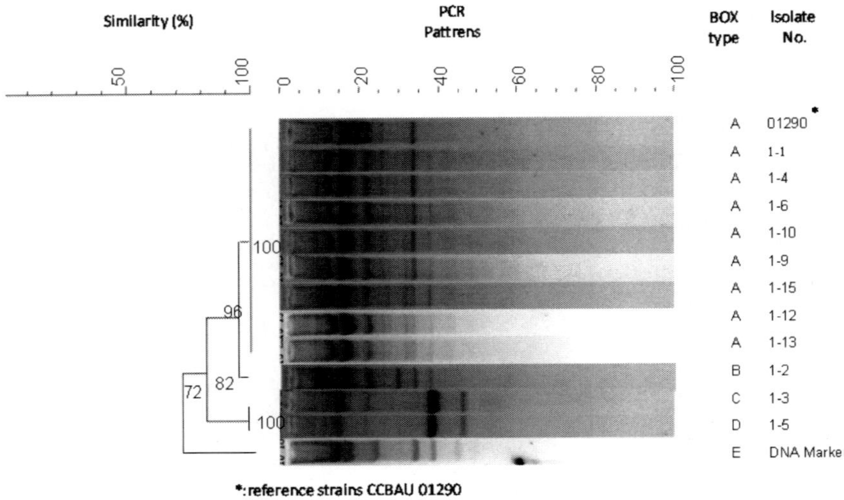

Figure 3. Example of BOX-PCR fingerprinting showing the comparison of nodule isolates and the inoculated strain (CCBAU 01290) that was used to estimate the nodule occupancy of the inoculant. Dice coefficient (Optimization: 5.00%, Tolerance: 5.0%-10%) was calculated with the BOX PCR patterns and UPGMA (unweighted pair group method with arithmetic mean) method was used to construct the dendrogram. The isolates sharing 100% similarity were identified as the same BOX type (strain). Box type A was identified as the inoculated strain CCBAU 01290. BOX types B, C, and D represent indigenous rhizobia in the soil. BOX type E represent DNA marker (Molecular weight range from 2000bp to 200bp).

Most of the low and intermediately effective strains showed relatively low nodulation occupancy or competition in soil, especially the strain CCBAU 13012 that had the smallest occupation of 25%.

Biomass accumulation In this test, the dry weights of alfalfa plants inoculated with test strains were compared.

The results demonstrated clearly that only the four effective strains of the group 1 significantly enhanced the shoot weights of alfalfa in comparison with the blank control.

Therefore, these four strains were both high efficient and high competitive rhizobia in association with alfalfa in soil of Inner Mongolia.

Field experiments The data summarized in Table 2 demonstrated that the four strains of group 1 showed high competence and high efficiency in greenhouse experiments also significantly enhanced the biomass accumulation of alfalfa in fields, in comparison with the no inoculated control and with the inefficient strain CCBAU 13012.

Nitrogen Fixation Variation among the Alfalfa Cultivars and Rhizobial Isolates

Measurement of plant growth showed that the height of 'Haygrazer' and 'Aohan' was significantly increased (P=0.05) by inoculation with CCBAU 01290 and CCBAU83164 (Figure 5A); the shoot weight of 'Haygrazer' and 'Vector' was significantly increased by inoculation with CCBAU01199 and CCBAU01290 (Figure 5B); total plant nitrogen content was significantly increased by inoculation with CCBAU01199 and CCBAU01290 in all three alfalfa cultivars (Figure 5C); total plant phosphorus content varied with the different combinations (Figure 5D). These results demonstrated that distinct combinations of rhizobial strains and alfalfa cultivars affected plant growth differently, which confirmed earlier findings [33]. Inoculation with the four highly effective strains significantly increased the nodule numbers (31.58% to 158.82%) on tested cultivars; while low effective strain CCBAU13012 showed negative effects (Figure 5E). The nodule occupation varied (22% to 54%) for the five inoculated strains on three cultivars (Table 3). It seems that the strains CCBAU01314 and CCBAU01199 are highly competitive microsymbionts for all three cultivars, which have an average occupation of 50% and 45%, while these values were 30.6% to 38.6% for the other three strains. The inoculation of **exogenous** rhizobia changed the composition and relative abundance of the native rhizobia in the nodules (Table 3), but did not affect the absolute numbers of the nodules occupied by the native rhizobia in some cases, considering the significant increase of the total nodule numbers by the inoculation (Table 3).

These results imply that the inoculated strains can increase the number of nodules formed by the native strains, which suggests not only competition, but also cooperation. It has been shown that alfalfa cultivars affected the *E. meliloti* populations [40]. Our results demonstrated that the inoculation of exogenous rhizobia also modified the diversity of native rhizobia in the nodules (Figure 6).

The occurrence of new BOX-types (CCBAU01290 vs. 'Haygrazer'; CCBAU01199 vs. 'Vector' etc.) and the disappearance of some BOX types (CCBAU13012 vs. 'Aohan') in inoculated treatments indicated that the inoculants might help or inhibit the infection of some native rhizobia.

Although the mechanism of the positive interaction between the inoculated and native rhizobia is not clear, it could be concluded that the ability of nodulation and competition were affected by both the host plants and the rhizobia at the level of plant cultivars and rhizobial strains.

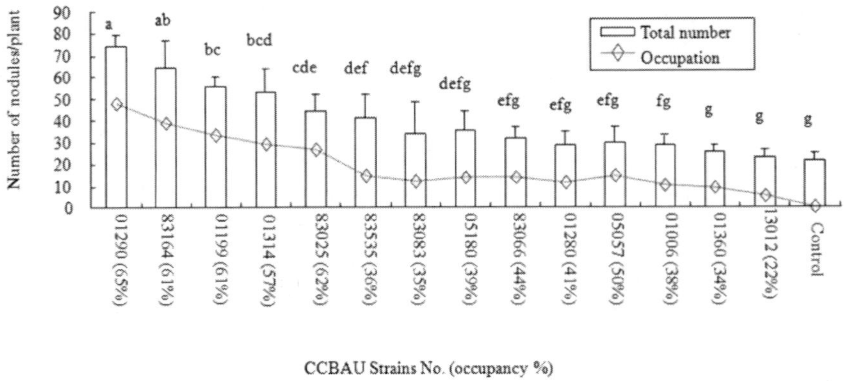

Figure 4. Results of competitive analysis of alfalfa rhizobial in natural soils. A: total nodule numbers/occupation of inoculated strains, R-square = 0.812206 (P < 0.0001). B: fresh weight of alfalfa plants inoculated with test strains, R2 = 0.689239 (P > F = 0.0002). Means marked with the same letters are not significantly different in the statistic analysis with Duncan's Multiple Range Test in ANOVA program of SAS software package.

Table 2. Inoculation effects of rhizobial strains on alfalfa biomass accumulation

Inoculant	Average dry weight (g plant^{-1})*	Std. error
CCBAU 01199	4.114286^a	0.774389
CCBAU 01290	4.156857^a	0.768219
CCBAU 01314	3.525200^a	0.739870
CCBAU 83164	3.566000^a	0.622855
CCBAU 13012	3.171667^b	0.475387
Blank control	3.030000$_b$	0.267836

*Numbers marked with different letters were significantly different in statistical analysis.

The principle component analysis (PCA) showed that *E. meliloti* CCBAU01199 and CCBAU01290 had broad-spectrum effectiveness, which resulted in better plant characteristics for all three cultivars; strains CCBAU01314 and CCBAU83164 had less effectiveness; while the low effective strain, CCBAU13012, ranked with the blank control (Figure 7).

Our results confirmed a preference between rhizobial strains and alfalfa cultivars, and revealed the presence of wide spectrum rhizobial strains (CCBAU01290 and CCBAU01199) which showed better nodulation and growth promotion on different alfalfa cultivars.

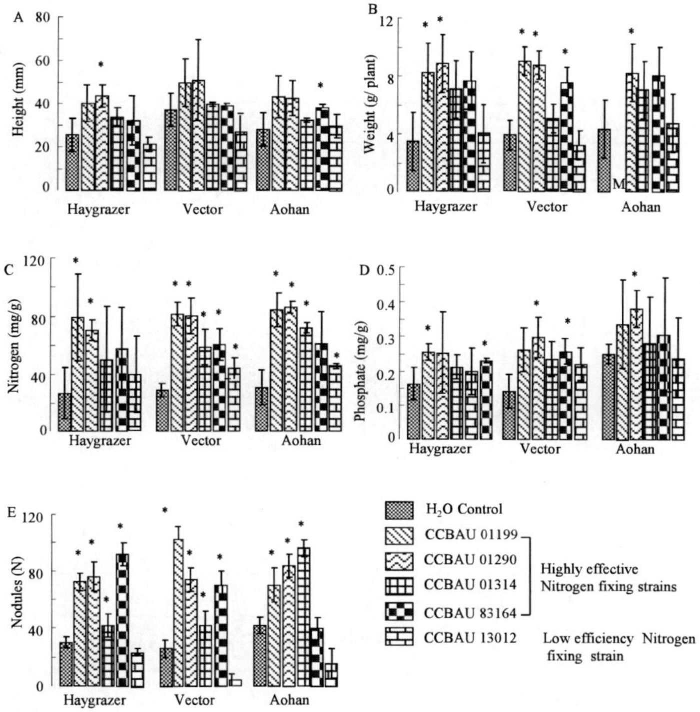

M: indicates data missing.

Figure 5. Comparison of the biomass and symbiotic characteristics obtained with different combinations of rhizobial strains and three alfalfa cultivars. The cultivars tested were Haygrazer, Vector and Aohan. The statistical analysis was performed with ANOVA at P=0.05. (A) Height of shoots of alfalfa in different treatments (inoculation vs cultivar). (B) Dry weight of alfalfa in different treatments. (C) Total phosphate content of alfalfa with each combination. (D) Total nitrogen content of alfalfa with each combination. (E) Nodule numbers per plant for different treatments.

Meanwhile, the inoculated strains and the native rhizobia may compete for nodule occupancy, but cooperated to increase the nodule numbers.

Yields of Alfalfa and Smooth Brome Intercropping and Monocropping

Plant height and dry weight was recorded during the time taken for two mows (Figure 8). In the first mow the plant height of alfalfa and smooth brome did not show any significant difference in intercropping between inoculation/ non-inoculation treatments.

Table 3. Characterization of BOX types and the nodule occupancy of rhizobia isolated from root nodules of different alfalfa cultivars

BOX-Fingerprinting		Nodulation Occupation				
Bacterial Strains	PCR patterns	Alfalfa Cultivars	Inoculations	NI[c]	NT[d]	NN[e]
		Haygrazer	Blank[b]	-	4	50
CCBAU01199			CCBAU 01199	25	5	25
CCBAU01290			CCBAU 01290	14	11	36
CCBAU01314	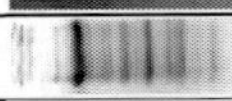		CCBAU 01314	27	4	23
CCBAU83164			CCBAU 83164	22	9	28
CCBAU13012			CCBAU 13012	11	9	39
		Vector	Blank	-	4	50
			CCBAU 01199	18	10	32
Native 1[a]			CCBAU 01290	26	6	24
Native 2	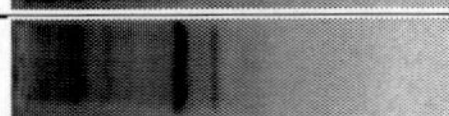		CCBAU 01314	25	4	25
Native 3	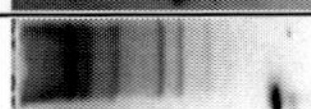		CCBAU 83164	17	6	33
Native 4	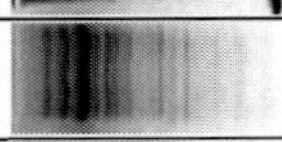		CCBAU 13012	20	5	30
Native 5		Aohan	Blank	-	4	50
Native 6			CCBAU 01199	25	8	25
Native 7	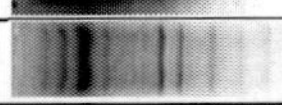		CCBAU 01290	18	6	32
Native 8			CCBAU 01314	23	8	27
Native 9	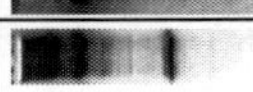		CCBAU 83164	18	5	32
Native 10	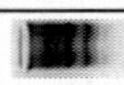		CCBAU 13012	15	3	35

[a]Native 1 to 10 BOX types were identified among native rhizobia.

[b]Blank indicates the control (H_2O) treatment.

[c]NI represents the number of isolates which have BOX types corresponding to the inoculated strains.

[d]NT indicated the number of the BOX types represents native rhizobia.

[e]NN indicated the number of native rhizobia among the isolates.

'-'. not detected.

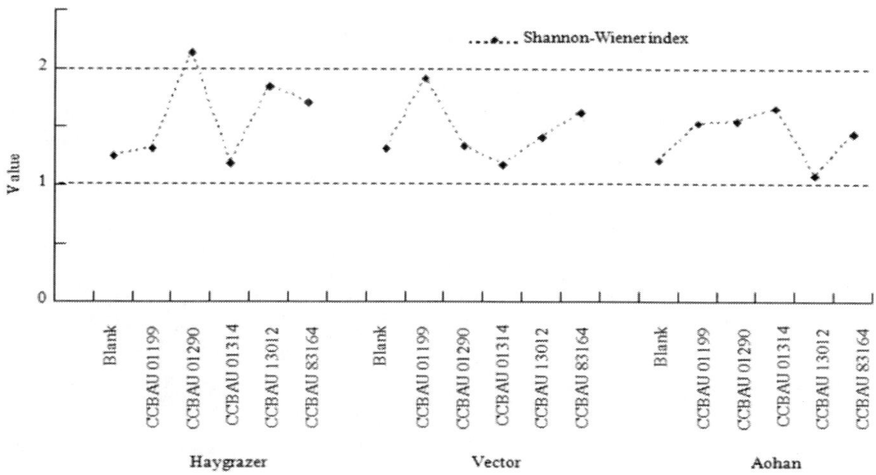

Figure 6. Biodiversity of native rhizobia in alfalfa nodules affected by inoculation with rhizobial strains. Shannon-Wiener, where s represents number of BOX-type, and pi means the proportion of individuals of the total sample belonging to the ith BOX type.

Monocropping alfalfa after inoculation with rhizobia showed a significant increase in the height of the plant, further confirming the effectiveness of the rhizobia.

In the monocropping of smooth brome there was no significant difference between the rhizobial inoculation and non-inoculation treatments. In the second mow, while the average plant height was higher than in the first mow, there was no significant difference in the height of alfalfa and smooth brome between intercropping and monocropping, with or without inoculation.

The dry weight of the alfalfa from AB_R (Alfalfa intercropping with smooth brome with rhizobia inoculation) was significantly heavier than that in AB_ck (Alfalfa intercropping with smooth brome without rhizobia inoculation) in the first mow (Figure 8). The intercropping partner smooth brome produced a similar yield in both AB_R and AB_ck. Monocropping alfalfa or monocropping smooth brome with inoculation (MA_R or MB_R) produced a higher biomass compared to MA_ck (monocropping alfalfa without rhizobial inoculation) and MB_ck (monocropping smooth brome without rhizobia inoculation), respectively. The difference in yield of alfalfa and smooth brome was even greater in the second mow. Both alfalfa and smooth brome from AB_R were significantly heavier than from AB_ck and the weight of alfalfa from MA_R was significantly heavier than that from MA_ck. The weight of smooth brome in monocropping was not significantly different between inoculation and non-inoculation treatments.

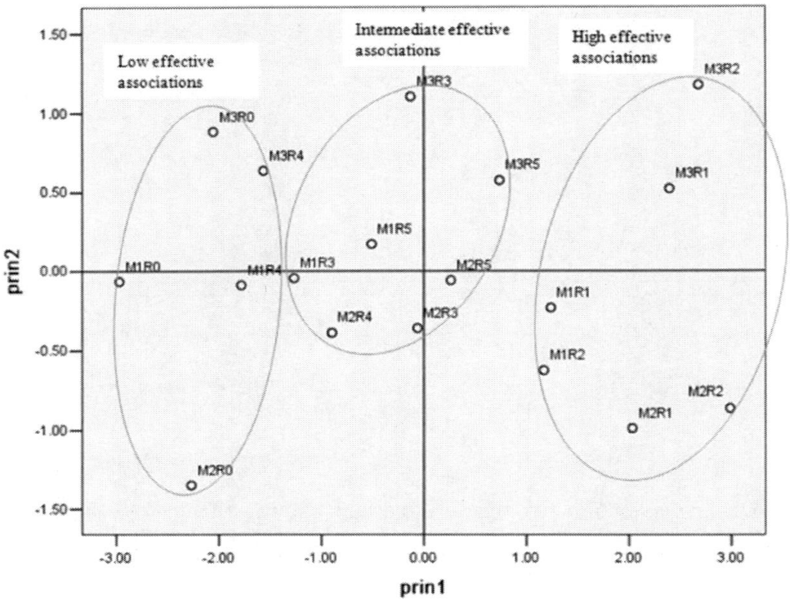

Figure 7. Principle component analysis (PCA) of different combinations between rhizobial isolates and alfalfa cultivars PCA conducted with SAS (SAS v8 inc.) using all the bioassay data (plant height, plant dry weight, plant nitrogen, plant phosphorus, number of nodule, inoculums occupation, and native rhizobial diversity). M=M. sativa cultivar; M1: Haygrazer; M2: Vector; M3: Aohan. R=rhizobial; R0: control (inoculated with H2O), R1 to R5: inoculated with CCBAU01199, CCBAU01290, CCBAU01314, CCBAU13012 and CCBAU83164. prin1 and prin2 indicate the first and second principal components.

The total number of nodules from alfalfa was also recorded. The number of nodules from AB_R was significantly higher than in AB_ck, while the greatest difference in nodule number was in comparing MB_R with MB_ck. Overall the average number of nodules was much higher in the intercropping system than in the monocropping system. Similar results were also found in the second mow.

Dynamic Biodiversity of Microorganisms

The agricultural practices of intercropping/monocropping and rhizobial inoculation altered the diversity of soil microorganisms and this was shown by T-RFLP (Figure 9A). The microbial composition also varied in the 6 months after inoculation (Figure 9B).

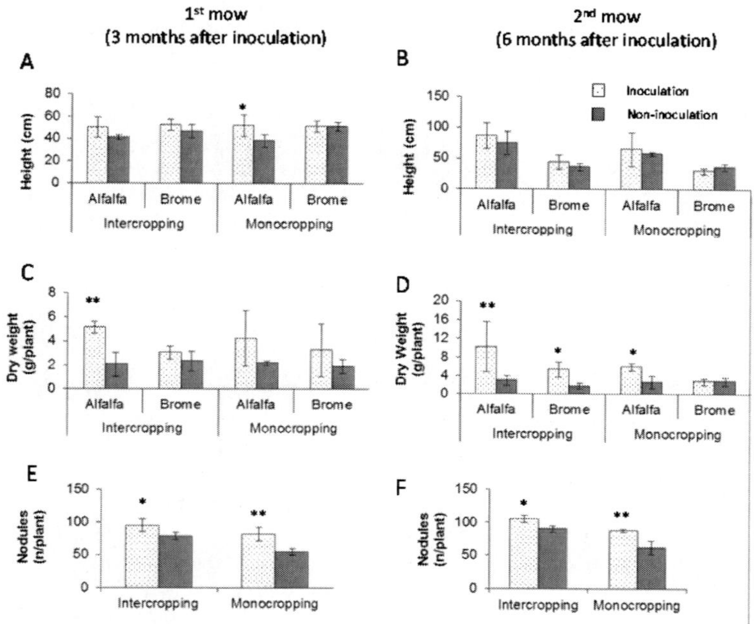

The bars means represented standard error, and *for significant level P=0.05, **for P=0.01.

Figure 8. Biomass and nodulation measured in alfalfa and smooth brome intercropping and monocropping. The first mow time, also 3 months after inoculation, the means of 6 replicates of plant height (A), dry weight (C), and the means of 3 replicates of nodule numbers (E) were presented. The second mow time, also 6 months after inoculation, the means of 3 replicates of plant height (B), dry weight (D), and nodule number (F) were recorded.

In the first months after inoculation, in the alfalfa and smooth brome intercropping with rhizobia inoculation (AB_R), the diversity was not changed, although the diversity from non-inoculation treatment (AB_ck) was sharply decreased. In the monocropping alfalfa with rhizobia inoculation (MA_R), the microbial diversity was significantly increased while for the non-inoculation treatment (MA_ck) there was only a slight increase, less than that of MA_R. In monocropping smooth brome with rhizobia inoculation system (MB_R), even though the rhizobia could form symbiosis with smooth brome, the diversity in MB_R reached a peak a month after inoculation with rhizobia. On the other hand, non-inoculation treatments (MB_ck) also showed decreasing diversity similar to that seen in AB_ck. In the first month, without inoculation, MB_ck and AB_ck showed a decrease in diversity, with AB_ck being lower.

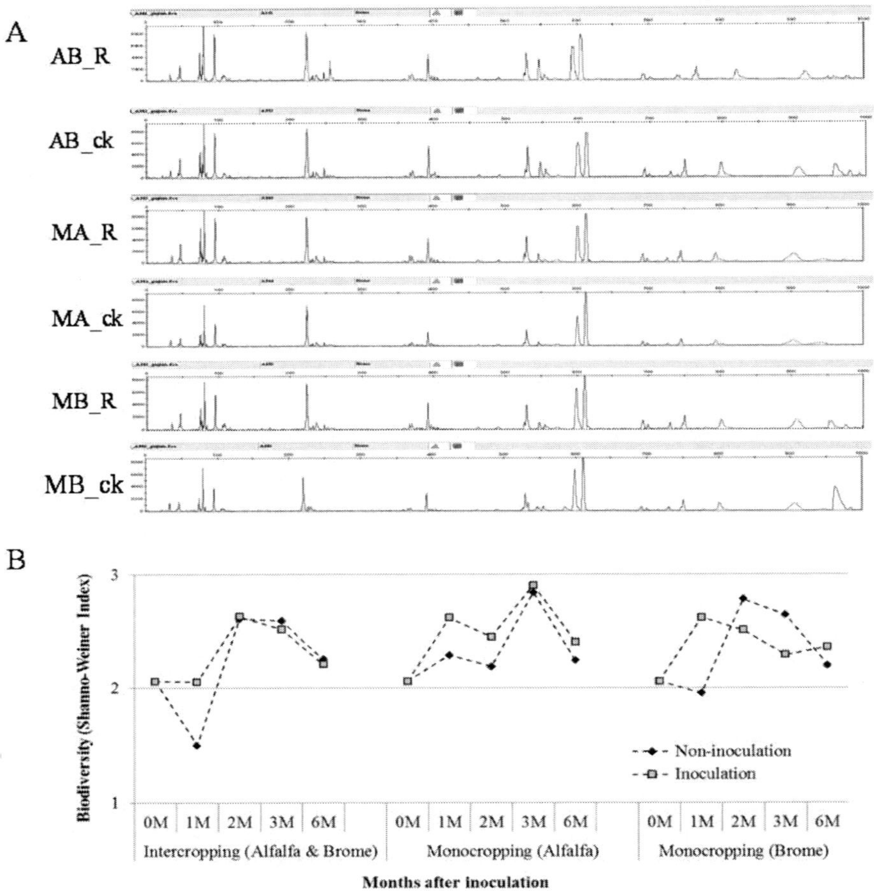

Figure 9. T-RFLP profiles and diversity of microorganism in soil. (A) Examples of TFs profile soil was collected at 2MAIs. (B) Dynamical biodiversity of microorganism communities alerted by inoculated rhizobia in alfalfa-brome grass intercropping system. The biodiversity was revealed by Shannon-Weiner index (described in Method).

The second month after inoculation, the biodiversity of microorganisms from intercropping peaked rapidly in both rhizobial inoculation (AB_R) and non-inoculation treatments (AB_ck). Such increasing diversity was not seen in monocropping alfalfa, but in both rhizobia inoculation (MA_R) and non-inoculation treatments (MA_ck), there was a slight decrease. In non-inoculated monocropping smooth brome (MB_ck), after an initial decrease, the diversity boomed in the second month, while the inoculation treatment (MB_R) decreased gradually. Additionally, monocropping smooth brome was the only system in which diversity of the inoculation treatment (MB_R) was lower than that of the non-inoculation treatment (MB_ck), presumably due to inoculation missing the host plants.

In the third and sixth months after inoculation, the diversity of the micro-community for AB_R and AB_ck was reduced, although both produced a similar and higher level of diversity than the original soil before planting. Diversity in MA_R and MA_ck reached a peak 3 months after inoculation, and then sharply decreased to close to the original soil diversity before planting.

An unusual observation was seen for MB_R which decreased from 2 to 3 months after inoculation but was slightly increased at 6 months. Compared to MB_ck, the soil diversity from MB_R was still low at 3 months. This situation was reversed at 6 months after inoculation, as MB_ck sharply decreased while MB_R increased.

To sum up the influence from the inoculation and cultivation practices, all treatments were compared to the original soil diversity prior to inoculation, noting increases and decreases in microbial diversity. In the intercropping system, such variation in both AB_R and AB_ck peaked at 2 months after inoculation, while monocropping alfalfa (MA_R and MA_ck) peaked 3 months after inoculation. The situation was very different in monocropping smooth brome. MB_R showed the highest diversity at 1 month after inoculation, while for MB_ck the highest level was at 2 months after inoculation. Also, higher diversity was observed in the MB_ck treatment rather than MB_R.

Conclusion

The aim of inoculation of plants with microorganisms is to offer the plants some benefits, such as improve their growth or health. In the case of rhizobial, inoculation is for helping the legumes to fix more nitrogen and hitherto to have a greater productivity. The success of inoculation of rhizobia depends on several factors, but the basal ones are the effectiveness related to the nodulation and nitrogen-fixation abilities and the competitiveness with indigenous bacterial populations [41]. In general, screening of strains with high effectiveness and competitiveness as inoculants is performed with two-step strategy considering that good correlations between nodulation parameters and plant productivity are always obtained under N-deficient conditions [42], but some strains may be not good to be applied in natural soils due to the failure to compete with indigenous rhizobia [43].

Aiming at selecting rhizobial inoculants for alfalfa grown in Inner Mongolia, a two-step protocol was carried out in this study to obtain the best strains and the results were confirmed with field experiments. Firstly, the effectiveness of all the strains was tested in greenhouse with the vermiculite as the nutritionfree media by combined analysis of dry weight of shoots, total number of nodules and activity of nitrogenase estimated by acetylene reducing activity. Secondly, the high effective strains selected in step I and other randomly chosen strains were employed for testing competition and effectiveness in natural soil collected from Inner Mongolia. The performance of competition of the inoculated strains with indigenous rhizobia in the soil was presented as nodule number and occupancy estimated with BOX PCR fingerprinting, and the effectiveness of inoculated strains in soil was estimated from the increase of alfalfa biomass. With this protocol, four most effective strains were obtained among the 96 test strains in the first step and they were also the most competitive and effective strains in the greenhouse experiments using natural soil of Inner Mongolia. All the four most effective strains were isolated from Inner Mongolia and North Xinjiang (eco-region I). This relationship between the nodulation parameters and geographic origin of the strains confirmed our previous observation that the symbiosis between rhizobia and legumes is a result of interactions among the partners and environmental factors [4].

Recently, Lesueur et al. (2001) selected from 446 isolates 6 powerful strains nodulating with *Calliandra calothyrus* [44]. These 6 strains were respectively isolated from Kenya, Costa Rica, and New-Caledonia, distributing in Africa and South America.

However, this screen was performed only in sand and the soil conditions were not considered. Previously, we have observed that *Bradyrhizobium* sp. was predominant in nodules of common bean (*Phaseolus vulgaris*) grown in an acid soil [4]; *Sinorhizobium fredii* associated with soybean and *S. meliloti* associated with alfalfa [2] in Xinjiang (alkaline-saline soils) were subpopulations different from those reported in other regions, where the genotypes of *Rhizobium leguminosarum* strains associated with faba bean (*Vicia faba*) showed clear biogeography in three eco-regions [45]. All the previous reports and the results in the present study demonstrated that the selection or introduction of rhizobial inoculants should consider the soil conditions or ecoregions. Detection of target microorganism in environmental samples is a universal demand in studies on interrelations between microbe-

microbe, microbe-materials and microbe-plant and animal, and one of the important models is rhizobia-legume interaction.

Some more sensitive and distinctive methods, such as BOX-A1R-PCR, and T-RFLP were introduced into ecological studies [27]. The latter techniques could detect the target bacteria based the original genetics, without any modified genes, and unforeseeable interference. In the present study, BOX-PCR functions very well in the differentiation of inoculated strains from the indigenous strains, although the alfalfa rhizobial have very similar phylogenies. Under the described procedure, we could obtain highly repeatable results. In addition, all the four best strains selected in the first step, estimated with principal component analysis, were exactly the great competitive ones in the second step and the best ones in field experiments. This relationship revealed the great value of principal component analysis and it might significantly enhance the efficiency of rhizobial inoculants selection.

To sum up, the comprehensive procedure employed in the present study was very effective in scanning large number of rhizobial strains for selection of inoculants. It was clear that the alfalfa rhizobial strains had very different capacities in nodulation, nitrogen-fixation and competition, and the most effective and competitive strains were all found among the test strains isolated from the same eco-regions of the target alfalfa cultivar, indicating the necessity of selection of inoculants according the eco-regions or ecotypes of the target plants. Based upon the results, we selected the 4 strains CCBAU 01199, CCBAU 01290, CCBAU 01314 and CCBAU 83164 as high effective and high competitive inoculants for alfalfa grown in eco-region I of China.

The impact of inoculation on plant growth and on nodulation of native rhizobia was analyzed using combinations of five *Sinorhizobium meliloti* strains and three alfalfa cultivars. The strains CCBAU01199 and CCBAU 01290 showed a broad host-spectrum, improving the growth and nitrogen fixation in all three alfalfa cultivars (32-156%), with nodule occupation 22-54%. The diversity of native rhizobia in the alfalfa nodules may be reduced or increased by inoculation, depending on both the cultivar and the strain. These results confirmed a preference between the rhizobial strains and the plant cultivars, and revealed that the inoculants and native rhizobia competed for occupancy of nodules and cooperated to enhance the nodule numbers.

Our further results showed that rhizobial inoculation and intercropping both promoted nitrogen fixation. The diversity of soil microorganisms varied with each treatment, however, after 6 months the microbial composition in each cropping system returned to normal.

Rhizobia are one of most studied and well best characterized soil bacteria. It has been proven as effective management to replace chemical fertilizer in legume crops and by means of nitrogen fixation, atmospheric nitrogen can be recycled for the next non-legume crops. The benefits of adding rhizobia were amplified in legume and non-legume intercropping. Benefits included improving soil structure, breaking pest cycles, decreasing disease infestation, and reducing greenhouse gas emissions. Rhizobia bacteria also severed served as PGPR (Plant growth promoting rhizo-bacteria) by releasing secondary metabolites which act as plant growth regulators / phytohormones / biologically active substances, as well as enhancing the plant nutrition.

Intercropping legume and non-legume also enhanced the land equivalent ratio (LER) as chemical fertilizer tends to suppress nitrogen fixation and greater efficiency without fertilizer application was observed. The rhizobia, legume and non-legume system also accounted to minimize the soil fertility depletion of soil fertility.

ACKNOWLEDGMENTS

The authors would like to thank Dr. Heather McCafferty in Hawaii Agriculture Research Center for her expert comments as well as English writing.

REFERENCES

[1] Allen, O. N. and Allen, E. (1981). *The Leguminosae: A Source Book of Characteristics, Uses, and Nodulation*. University of Wisconsin Press, Madison, WI.

[2] Yan, A. M., Wang, E. T., Kan, F. L., Tan, Z. Y., Sui, X. H., Reinhold-Hurek, B., Chen, W. X. (2000). *Sinorhizobium meliloti* associated with *Medicago sativa* and *Melilotus* spp. in arid saline soils in Xinjiang, China. *Int. J. Syst. Evol. Microbiol.*, 50: 1887-1891.

[3] Kan, F. L., Chen, Z. Y., Wang, E. T., Tian, C. F., Sui, X. H., Chen, W. X. (2007). Characterization of symbiotic and endophytic bacteria isolated from root nodules of herbaceous legumes grown in Qinghai-Tibet plateau and in other zones of China. *Arch. Microbiol.*, 188: 103-115.

[4] Han, S. Z., Wang, E. T., Chen, W. X. (2005). Diverse bacteria isolated from root nodules of *Phaseolus vulgaris* and species within the genera *Campylotropis* and *Cassia* grown in China. *Syst. Appl. Microbiol.*, 28: 265-276.

[5] Zeng, Z. H., Sui, X. H., Hu, Y. G., Chen, D. M., Chen, W. X., Gao, R. L. (2004). Screening of high-effective *Sinorhizobium meliloti* strains for *Medicago sativa* cultivars and their field inoculation. *Cao Ye Xue Bao* 13: 95-100 [in Chinese].

[6] Zhao, S. H., Wang, L. F., Chen, W. X. (2005). Study on the affect of rhizobia inoculated on alfalfa cultivars. *Hebei Ke Xue Yuan Xue Bao* 22: 63-66 [in Chinese].

[7] Yang, Q. J., Zhou, H., Lu, P., Qi, L. (2006). Study on Rhizobium inoculation and seed pill clothing of alfalfa. *Chinese Agri. Sci. Bull.*, 22: 52-55.

[8] Lu, X. S. (1998). Winter hardiness evaluation in Chinese alfalfa cultivars. *Grassland of China*. 3: 1-5 [in Chinese].

[9] Tsubo, M., Walker, S., Ogindo, H. O. (2005). A simulation model of cereal-legume intercropping systems for semi-arid regions: I. Model development. *Field Crop Res.* 93: 10-22.

[10] Willems, A. (2006). The taxonomy of rhizobia: an overview. *Plant Soil* 287: 3-14.

[11] Cheng, Q. (2008). Perspectives in biological nitrogen fixation research. *J. Integr. Plant Biol.* 50: 786-798.

[12] Krishna, K. R. (2010). *Agroecosystems of South India: Nutrient dynamics ecology and productivity*. Brown Walker press, Boca Raton, Florida US.

[13] Ullah, M. A. (2010). Forage production in Panicum grass-legumes intercropping by combining geometrical configuration, inoculation and fertilizer under rainfed contiditions. *Thesis, Kassel University*, Germany.

[14] Chianu, J. N., Nkonya, E. M., Mairura, F. S., Chianu, J. N., Akinifesi, F. K. (2011). Biolgocial nitrogen fixation and socioeconomic factors for legume production in sub-Saharan Africa. *Agron. Sustain. Dev.* 31: 139-154.

[15] Wilson, P. W. and Burris, R. H. (1953). Biological nitrogen fixation-a reappraisal. *Annu. Rev. Microbiol.* 7: 415-432.

[16] Cytryn, E., Levkovitch, I., Negreanu, Y., Dowd, S., Frenk, S., Silber, A. (2012). Impact of short-term acidification on nitrification and nitrifying bacterial community dynamics in soilless cultivation media. *Appl. Environ. Microbiol.* 78: 6576-6582.

[17] Dudman, W. F. and Belbin, L. (1988). Numerical taxonomic analysis of some strains of Rhizobium spp. that uses a qualitative coding of immunodiffusion reactions. *Appl. Environ. Microbiol*. 54: 1825-1830.

[18] Schmidt, E. L., Bankole, R. O., Bohlool, B. B. (1968). Fluorescent antibody approach to the study of rhizobia in soil. *J. Bacteriol*. 95: 1987-1992.

[19] Berger, J. A., May, S. N., Berger, L. R., Bohlool, B. B. (1979). Colorimetric enzyme-linked immunosorbent assay for the identification of strains of Rhizobium in culture and in the nodules of lentils. *Appl. Environ. Microbiol.* 37: 642-646.

[20] De Maagd, R., van Rossum, C., Lugtenberg, B. J. (1988). Recognition of individual strains of fast-growing rhizobia by using profiles of membrane proteins and lipopolysaccharides. *J. Bacteriol*. 170: 3782-3785.

[21] Moreira, F. M. S., Gillis, M., Pot, B., Kersters, K., Franco, A. A. (1993). Characterization of rhizobia isolated from different divergence groups of tropical *Leguminosae* by comparative polyacrylamide gel electrophoresis of their total proteins. *Syst. Appl. Microbiol*. 16: 135-146.

[22] Young, J. P. W. and Haukka, K. E. (1996). Diversity and phylogeny of rhizobia. *New Phytol*. 133: 87-94.

[23] Giller, K. E., McGrath, S. P., Hirsch, P. R. (1989). Absence of nitrogen fixation in clover grown on soil subject to long-term contamination with heavy metals is due to survival of only ineffective *Rhizobium*. *Soil Biol. Biochem*. 21: 841-848.

[24] Selander, R. K., Caugant, D. A., Ochman, H., Musser, J. M., Gilmour, M. N., Whittam, T. S. (1986). Methods of multilocus enzyme electrophoresis for bacterial population genetics and systematics. *Appl. Environ. Microbiol.* 51: 873-884.

[25] Williams, J. G., Kubelik, A. R., Livak, K. J., Rafalski, J. A., Tingey, S. V. (1990). DNA polymorphisms amplified by arbitrary primers are useful as genetic markers. *Nucleic Acids Res*. 18: 6531-6535.

[26] Vos, P., Hogers, R., Bleeker, M., Reijans, M., van de Lee, T., Hornes, M., Frijters, A., Pot, J., Peleman, J., Kuiper, M., et al. (1995). AFLP: a new technique for DNA fingerprinting. *Nucleic Acids Res*. 23: 4407-4414.

[27] Laguerre, G. L., Louvrier, P., Allard, M.-R., Amarger, N. L. (2003). Compatibility of rhizobial genotypes within natural populations of

Rhizobium leguminosarum biovar *viciae* for nodulation of host legumes. *Appl. Environ. Microbiol.* 69: 2276-2283.

[28] Peix, A., Velazquez, E., Martinez-Molina, E. (2007). Molecular methods for biodiversity analysis of phosphate solubilizing microorganisms (PSM). In: Velazquez, E., Rodriguez-Barrueco, C. (ed), *First international meeting on microbial phosphate solubilization*. Springer, Netherlands. pp. 97-100.

[29] Børresen, A. L., Hovig, E., Brøgger, A. (1988). Detection of base mutations in genomic DNA using denaturing gradient gel electrophoresis (DGGE) followed by transfer and hybridization with gene-specific probes. *Mutat. Res.* 202: 77-83.

[30] Maidak, B. L., Cole, J. R., Parker, C. T. J., Garrity, G. M., Larsen, N., Li, B., Lilburn, T. G., McCaughey, M. J., Olsen, G. J., Overbeek, R., Pramanik, S., Schmidt, T. M., Tiedje, J. M., Woese, C. R. (1999). A new version of the RDP (Ribosomal Database Project). *Nucleic Acids Res.* 27: 171-173.

[31] Sun, Y. M., Zhang, N. N., Wang, E. T., Yuan, H. L., Yang, J. S., Chen, W. X. (2009). Influence of intercropping and intercropping plus rhizobial inoculation on microbial activity and community composition in rhizosphere of alfalfa (*Medicago sativa* L.) and Siberian wild rye (*Elymus sibiricus* L.). *FEMS Microbiol. Ecol.* 70: 218-226.

[32] Vincent, J. M. (1970). A Manual for the Practical Study of the Root-Nodule Bacteria. *IBP Handbook 15*, Blackwell Scientific Publications, Oxford.

[33] Bromfield, E. S. P. (1984). Variation in preference for *Rhizobium meliloti* within and between *Medicago sativa* cultivars grown in soil. *Appl. Environ. Microbiol.* 48: 123-1236.

[34] Jia, R. Z., Tian, C. F., Man, C. X., Wang, E. T., Chen, W. X. (2008). Screening of high effective alfalfa rhizobial strains with a comprehensive protocol. *Ann. Microbiol.* 58: 731-739.

[35] Staal, M., Lintel-Hekkert, S. T., Harren, F., Stal, L. (2001). Nitrogenase activity in cyanobacteria measured by the acetylene reduction assay: a comparison between batch incubation and online monitoring. *Environ. Microbiol.* 3: 343-351.

[36] Jia, R. Z., Wang, E. T., Liu, J. H., Li, Y., Gu, J., Yuan, H. L., Chen, W. X. (2013). Effectiveness of different *Ensifer meliloti* strain-alfalfa cultivar combinations and their influence on nodulation of native rhizobia. *Soil Biol. Biochem.* 57: 960-963.

[37] Jia, R. Z. (2009) Evaluation and Application highly effective rhizobia-alfafa-bromus intercropping system in agriculture. *China Agriculture University* [Thesis in Chinese].

[38] Bremner, J. M. (1965). Total nitogen. *Methods of soil analysis, Part 2. Chemcial and microbiological properties.* Black, C. A., Evans, D. D., Ensminger, L. E., White, J. L., and Clard, F. E. Madison, Wisconsin, American Society of Agronomy Inc. 9: 1149-1178.

[39] Cavell, A. J. (1955). The colorimetric determination of phosphorous in plant materials. *J. Sci. Food Agric.* 6, 479-481.

[40] Carelli, M., Gnocchi, S., Fancelli, S., Mengoni, A., Paffetti, D., Scotti, C., Bazzicalupo, M. (2000). Genetic diversity and dynamics of *Sinorhizobium meliloti* populations nodulating different alfalfa cultivars in Italian soils. *Appl. Environ. Microbiol.* 66, 4785-4789.

[41] Thies, J. E., Holmes, E M., Vachot, A. (2004). Application of molecular techniques to studies in *Rhizobium* ecology: a review. *Aust. J. Exp. Agr.,* 41: 299-319.

[42] Somasegaran, P. and Hoben, H. J. (1994) *Handbook of Rhizobia.* Springer Laboratory, New York.

[43] Streeter, J. G. (1994). Failure of inoculant rhizobia to overcome the dominance of indigenous strains for nodule formation. *Can. J. Microbiol.,* 40: 513-522.

[44] Lesueur, D., Ingleby, K., Odee, D., Chamberlain, J., Wilson, J., Manga, T. T., Sarrailh, J. M., Pottinger, A. (2001). Improvement of forage production in *Calliandra calothyrsus*: methodology for the identification of an effective inoculum containing Rhizobium strains and arbuscular mycorrhizal isolates. *J. Biotechnol.*, 91: 269-282.

[45] Tian, C. F., Wang, E. T., Han, T. X., Sui, X. H., Chen, W. X. (2007). Genetic diversity of rhizobia associated with *Vicia faba* in three ecological regions of China. *Arch. Microbiol.,* 188: 273-282.

In: Alfalfa
Editor: Marie L. Hirsh
ISBN: 978-1-62948-752-6

Chapter 3

IMPACT OF RODENT SPECIES ON ALFALFA CROPS: SOLVING A PROBLEM

_Goran I. Jokić_*
Institute for Pesticides and Environment Research,
Belgrade-Zemun, Serbia

ABSTRACT

Alfalfa belongs to a group of major forage crops due to its high productive potential and excellent quality of green biomass and grain, as well as a high content of proteins. The cropping practices for alfalfa, its coverage and height, particularly of seed crops, are conducive to formation of specific agroecological conditions that attract establishment and spread of rodent colonies. Optimal living in alfalfa is furthered by rodents' prominent adaptability to various conditions and types of open habitats, their high reproductive potential and polyphagous character.

Feeding is the primary cause of damage done by rodents in alfalfa crops from the time of sowing, when they consume seeds in the field, throughout the vegetation period when they feed on green biomass, until wintertime when they eat roots. Undoubtedly, they cause the most serious impact on seed crops by eating, injuring and collecting mature seeds. Considering the levels of damage in alfalfa fields, species of the genus Microtus (*Microtus arvalis, Microtus californicus*, *Microtus guentheri, Microtus montanus* and *Microtus pennsylvanicus*) are the most prominent

* Tel./Fax. +381 11 3076 133; 3076 136, jokicg@ptt.rs.

pests. In some alfalfa-growing regions, corresponding damage may be caused by *Thomomys bottae*, *Spermophilus beecheyi*, *Meriones unguiculatus* or *Apodemus sp.*

Modern alfalfa cropping practices include ways to reduce economic losses caused by feeding and other activities of rodents to their economically acceptable level. The many existing alternative control measures, including a topsoil cultivation, building of physical barriers, traps or stimulation of predators, mostly fail to produce desirable effects, so that rodenticide treatments remain the most effective and often most economical method of controlling rodents. The specific distribution and ecological characteristics of each rodent species, different cropping practices and legal limitation for treatments influence the manufacturing and application of various active ingredients and formulations intended for control of rodents in alfalfa fields.

This chapter will survey the most harmful rodent pests in alfalfa crops, their biology, characteristic signs of presence and damage that they cause. Besides some non-chemical methods of control, the available commercial rodenticides will be listed, and their application methods and efficacy described.

INTRODUCTION

Alfalfa (*Medicago sativa* L.) is grown worldwide as one of the most important forage crops (Štrbac et al., 1996) for all classes of livestock. It is also a crop that is important for advancement of other branches of the economy (Mueller et al., 2007). Its cultivation enriches the soil with biological nitrogen and improves soil texture (Rogers and Gorman, 1995).

During vegetation, alfalfa creates a very suitable agroecological environment for colonization, survival and reproduction of small mammals (Rogers and Gorman, 1995). Soil properties and cultural practices applied in alfalfa fields, as well as the crop life span are compatible with rodent needs. The highest yields can be achieved by growing alfalfa on high-quality, loose soil that is rich in organic matter (Orloff and Carlson, 1995). Alfalfa is grown in monoculture and its life span is usually up to 5 years (Mueller et al., 2007). On the other hand, high adaptability to various conditions and types of habitat, ecological preferences regarding the main environmental factors, such as humidity and soil type, omnivorousness, high reproductive potential and cyclic overpopulation are the main features of the harmful rodent species which find optimal environment in alfalfa crops and create the main problem for their control.

Harmful rodents are able to cause significant yield and/or product quality reductions and often add to the shortening of productive life of alfalfa crops (Whisson and Salmon, 2007). Economic losses depend on the abundance and diversity of species. Several rodent species commonly inhabit an alfalfa crop simultaneously (Jokić et al., 2010; Jokić et al., 2012; Witmer and Singleton, 2010). Modern alfalfa cropping practices include methods for reducing to acceptable level the economic losses caused by feeding and other activities of rodents. Before implementing any specific damage management strategy, it is very important to determine the species living in a field. Knowing the characteristic signs of harmful rodents in alfalfa fields is a necessity. Various biotic and abiotic factors are crucial for choosing the best management strategy for rodent control in alfalfa crops (Singleton et al., 1999). Knowing the home range and seasonal activity period of a rodent species can help to determine the size of an area that should be put under intensive control, which may ultimately reduce the rate of reinvasion. Although a variety of methods have already been developed for reducing rodent damage in alfalfa crops (e.g., trapping, shooting, barriers, cultural practices, chemicals, etc.) there is still a need for new, environment-friendly and cost-effective approaches. The choice of a management strategy, especially when chemicals (rodenticides or fumigants) are planned to be used in the field, depends on local legislation, which is differnt across countries and regions. Also, it is vital to follow the labelled application instructions.

This chapter will survey the most harmful rodent pests and their main features (characteristic signs of presence, home range, activity, diet, etc.) in relation to alfalfa crops. Also, non-chemical and chemical control methods, their application and efficacy will be discussed.

SELECTING A FIELD FOR ALFALFA CROP

The importance of pre-crops, the condition in which a surrounding area is found (e.g., neglected), as well as total area, are factors that are often disregarded when choosing a field for growing alfalfa. To avoid damage from rodents, alfalfa is best to be sown in fields whose previous crop had been one that is least suitable for their settlement or reproduction, such as industrial crops which provide less favorable habitats. The possibility of choosing a site without surrounding areas that are suitable for rodent populations may significantly contribute to reducing losses in alfalfa production, especially in initial years of crop expoitation. The surrounding area (e.g., fence rows, roadsides, ditch banks) is a natural habitat for rodents in which they are normally found and from which they proceed to inhabit or invade arable land (Ružić, 1983).

Before setting up a crop, such a surrounding area, as well as its weed species and the height of plant cover, should be closely inspected. Large neglected areas abounding in weed vegetation that provides high-quality nutrition and protection from predators are probably the most unsuitable environment in which to plan sowing alfalfa.

Knowing the characteristics of pre-crops and surrounding areas, as well as the composition of populations of harmful rodents in them, makes it possible to choose optimal measures to protect a new alfalfa crop. Under the most unfavorable conditions, alfalfa protection from rodent pests needs to begin before the crop has actually been sown.

Regardless of choice, all protective measures should aim at reducing rodent numbers in the surrounding area and reducing the potential for their spreading into the alfalfa field. One of the ways to protect an alfalfa crop is to reduce the number of rodents along the field borders, close to its neglected neighboring area. Anyway, by cutting down the vegetative cover in all surrounding area it is possible to reduce the number of invading rodents.

Important Rodent Pests in Alfalfa Fields

It is almost impossible to find an alfalfa field free of at least one or several coexisting rodent species. Risks of yield loss may differ depending on the type of primary product collected from alfalfa fields, e.g., green foliage or seeds. To be able to protect an alfalfa crop, it is necessary to know the characteristics of some most important rodent species.

Based on what is currently known about harmful rodents in alfalfa fields around the world, voles, ground squirrels and gophers are considered the most important rodent species. Many other rodents, such as hamsters (*Cricetus* sp.), mice (*Apodemus* sp.) or zokors (*Myospalax* sp.) (Cash, 2009; Vukša et al., 2009) may also be sporadically important.

VOLES IN ALFALFA FIELDS

Voles can be found as a major pest species in almost all regions in which alfalfa is grown intensively.

In North America, *Microtus pennsylvanicus* (meadow mouse, field mouse) is the most widespread vole species (Maser and Storm, 1970). *Microtus montanus* (montane vole, Pahranagat Valley montane vole) (Sera and Early, 2003) and *Microtus californicus* (California meadow mouse) (Cudworth and Koprowski, 2010) are also commonly found as very harmful vole species in alfalfa crops. *Microtus arvalis* (common vole) is the most widespread Eurasian vole (Ricankova et al., 2007), while *Meriones unguiculatus* (Mongolian gerbil or Mongolian jird) naturally occurs in Eastern Asia (Gulotta, 1971). *Microtus guentheri* (Levant vole) (Wolf, 1977) can be found in the Mediterranean region and across the Middle East, from Turkey to Israel. Across the world, these vole species, as well as many of their subspecies have been recognized as very important pests in alfalfa crops. In addition, some other voles may also be relevant locally, such as *Microtus fortis* in China (Cash, 2009). Even though the body size of voles in alfalfa crops can vary depending on sex, age, geographic location, or subspecies, they are all small- or medium-sized. Body weight and lenght almost never exceed 90 g and 220 mm, respectively. They are active throughout the year and do not hibernate or aestivate. Daily activities depend on the climatic and habitat conditions or predators present in their environment. Most of them are considered nocturnal or crepuscular animals.

They live in colonies and are highly adaptable to various conditions and types of soil, preferring especially fertile soils (chernozem, loam, sandy loam) with a low level of underground water, a condition which is also desirable for an alfalfa crop (Cash, 2009). Soft soil is also suitable for rodents to dig their burrow systems in it. Many entrances are made to lead down into tunnels by oblique angles but rodents also make openings with nearly vertical entrances.

Even though alfalfa is a habitat that protects voles from predators for the most part of the year (Rogers and Gorman, 1995), they nevertheless spend most of their time during the day within their burrow systems. The number of openings depends on colony size and the type of soil. In most cases, some 10-20 exits can be located. Because they spend considerable time in their burrows, some voles are recognized as semi-fossorial (Batzli and Pitelka, 1970). The area populated by a colony, or containing a system of burrows, may reach several square meters (Ružić, 1983).

Clean ground runways, used to connect active holes, i.e., circular entrances of up to 3-6 cm in diameter (Ružić, 1983; Whisson and Salmon, 2007; Yavuz et al., 2010), are characteristic signs of vole presence. Their runways are especially easy to spot during wintertime before alfalfa has resumed growth. On dry soils or during arid and warm periods, runways are more difficult to detect. Shallow tunnels sunken into soil after abundant precipitation can also be observed. Fresh cut-outs of plant material and fecal pellets are found on runways, near or at the very entrances of active holes. Vegetation near the well-traveled runways may be clipped close to the ground.

Food supply, habitat and population density are very important aspects of the vole home range, which rarely exceeds a few hundred square metres (Banfield, 1974; Cash, 2009). During the vegetation period, when plant cover reaches 50 cm in height, the home range of common vole is several meters in diameter around active holes (Jacob and Hempel, 2003), and it has been reported to reach 145 m^2 (Mackin-Rogalska, 1981).

Voles are highly prolific mammals. Their peak breeding activity coincides with the culmination of alfalfa growth. In years with good climatic conditions, they are able to breed throughout the year. On the other hand, reproduction stops during unfavorable periods of the year. Females normally give birth 3-5 times a year with an average of 4-6 pups per brood. Because they are very prolific, there is a tendency of outbreaks. Cyclic overpopulation may occur at intervals of 2-5 years (Mendelssohn, 1972; Ružić, 1983; Whisson and Salmon, 2007). In years of overpopulation, the number of voles may reach thousands of animals per ha (Truszkowski, 1982). However, their life span in the wild is short, rarely exceeding one year.

Voles are herbivorous. During vegetation, they prefer to eat green foliage. In the fall, seeds make their favorable diet, while winter and spring are the seasons when they mostly consume alfalfa roots (Ružić, 1983; Whisson and Salmon, 2007; Yavuz et al., 2010). The considerable amount of energy that they spend during the day is recovered by very intensive feeding. However, their poor ingestion renders those energy sources quite low-efficient. They satisfy their energy needs by consuming large amounts of vegetative plant parts (Banin et al., 1993; Ostfeld, 1985). Some of them, such as meadow vole, have been observed to ingest close to 60% of their body weight in a day (Jackson, 1961). Also, but at a smaller percentage, their diet may include insects and fungi (Fogel and Trappe, 1978). Enhanced energy needs have been observed in pregnant and lactating females, i.e., up to 130 % higher than those of nonreproductive females (Migula, 1969).

GROUND SQUIRRELS IN ALFALFA FIELDS

Ground squirrel is considered a major rodent pest in alfalfa fields. Around the world, there are many species and subspecies of ground squirell. Among them, *Spermophilus beecheyi* (California ground squirrel) and *Spermophilus beldingi* (Belding ground squirrel) are considered the most significant squirell pests of alfalfa crops. In some regions of China, *Spermophilus dauricus* (Daurian ground squirrel) (Cash, 2009) has been found to cause great economic losses in alfalfa production.

S. beecheyi is widespread in California and Western Oregon, and throughout parts of Western Nevada and Washington (MacClintock, 1970), while *S. beldingi* naturally occurs in western parts of the United States, in California, Oregon, Nevada, Idaho and South-Eastern Utah (Jenkins and Eshelman, 1984). Generally, ground squirells are medium- or large-sized rodents. *S. beecheyi,* a large-sized species, has been found to reach 508 mm, and 700 g in body length and weight, respectively. In contrast, *S. beldingi* is a medium-sized species with the respective body length and weight of up to 300 mm and 300 g (Askham, 1994).

Generally, ground squirrels live in colonies and they are exceptionally adaptable to various climatic conditions. Depending on the climate in any particular region that they inhabit, they may hibernate or estivate during wintertime or some other unfavorable period of the year. Some young animals may skip hibernation in their first year (Evans and Holdenried, 1943; MacClintock, 1970).

Ground squirrels are also highly adaptable regarding habitat and type of nutrition. Sunny and open habitats rich in high-quality food, such as alfalfa plants, are suitable for settlement and reproduction of ground squirrels. They dig systems of burrows for many generations. A characteristic sign of their presence is a burrow system with many entrances that are always left open and never refilled with soil. Their openings are up to several cm in diameter, enough to fit their body size. Because *S. beldingi* is sometimes less colonial than *S. beecheyi*, its burrows can be more widely dispersed (Marsh, 1994). An area populated by its colony is often larger than the area of a *Thomomys bottae* colony, and its home range is up to 150 m in diameter. Ground squirrel is a terrestrial and semifossorial species.

Breeding occurs in the spring, a few weeks after hibernation or aestvation. Females have one litter annually and between five and eleven pups per litter. In the wild, they can live up to 6 years (MacClintock, 1970).

Even though they are highly omnivorous as a result of their changing energy needs over the year that are synchronized with vegetation periods, ground squirells are potentially important pests of alfalfa fields. They choose food of different energy and nutritive values to satisfy their varied needs over the year. In early spring, green foliage is the main source of energy, while nutrition during the period when they prepare for hibernation is mostly based on high-energy food, such as seeds and grain. Preparing for hibernation, the oldest animals nearly double their body weight by the middle of the year. They also eat roots, bulbs, fruits or insects (Marsh, 1998; Orloff and Carlson, 1995).

Pocket Gophers in Alfalfa Fields

Thomomys botae (valley pocket gopher) is considered the most important and most widely represented gopher species in alfalfa fields in North America (Miller, 1946; Orloff and Carlson, 1995). It can be found from Southern Oregon and Central Colorado to Southern Baja California and Central Mexico (Jones and Baxter, 2004). Besides *T. bottae*, *Geomys bursarius* and *Thomomys taploides*, and their numerous subspecies are also considered very important pocket gopher pests causing damage in alfalfa production. *G. bursaryus* naturally occurs in South Manitoba, Canada, and throughout the Central United States and southward to Texas (Matthew, 2011), while *T. talpoides* naturally occurs in the North American West and Midwest (Verts and Carraway, 1999). The size of pocket gophers can vary considerably depending on sex, age, geographic location or subspecies. Regarding morphological features, pocket gophers, *Thomomys* species in particular, are one of the most diverse mammals (Patton and Brylski, 1987). They are medium-sized and reach up to 30 cm in body length and 160 g in weight. Gophers of the *Geomys* genus are slightly larger (Matthew, 2011; Zimmerman, 1999).

Pocket gophers are solitary and territorial animals during most of the year, except their reproductive season (Orloff and Carlson, 1995). Their daily activity is particularly heightened at twilight and night-time (Benedix, 1994; Gettinger, 1984)

The shape of their mounds and underground tunnels greatly depend on the type of soil and variety of plant species. Mounds of fresh soil are the characteristic sign of pocket gophers present in an alfalfa field. The mounds are commonly crescent-shaped and appear at the end of short lateral tunnels (Orloff and Carlson, 1995). Tunnel length and spatial arrangement depend primarily on energy needs and available sources of high-quality food

(Scheffer, 1948; Thorne and Andersen, 1990). Generally, they build complicated burrow systems in arable soils. Tunnels of up to 8 cm in diameter are mostly dug at 30 cm depth, rarely deeper. Tunnel systems can take up an area of up to 100 m^2 (Whisson and Salmon, 2007). The animals spend most of their lives within their systems of tunnels, whose length can extent for tens or even hundreds of meters. They often use them for storing food and rarely go out of their shelter at daytime (Grzimek, 1990) as they are fossorial animals (Smith, 1998). Generally, the home range of pocket gophers is not especially wide and depends on various animal needs and available food.

Pocket gophers are active during the winter season. They can be inactive occasionally but without going into hibernation. Characteristic heaps of foods that they made even underneath the snow cover are one of the indicators of their presence and activity. Spring is the main breeding season. Females usually give birth to one litter annually, rarely two (Miller, 1946; Proulx and Cole, 2002). The average number of pups per litter is 3 to 7. Usually, the maximum life span in the wild of *T. bottae* and *T. talpoides* are 5 and 2 years, respectively (McMahon, 1999; Witmer and Engeman, 2007).

Pocket gophers are herbivores. Because of their specific mode of living, underground plant parts are the primary source of food, most particularly alfalfa roots. This type of nutrition has an exceptionally harmful effect on alfalfa crops. Injured plants mostly have no means of regenerating. Damage done by gophers to alfalfa plants is permanent. Apart from damage caused through feeding, pocket gophers also cause significant harm to alfalfa crops by digging their networks of tunnels and burying plants underneath soil mounds. Additionally, weed seeds buried underneath soil surface thus get into more favorable conditions for germination and emergence (Whisson and Salmon, 2007; Witmer and Engeman, 2007). This is particularly damaging to alfalfa seed crops. The fact that comparably larger specimens have been found in alfalfa than in other crops indicates the importance of alfalfa plants as a diet to pocket gophers (Patton and Brylski, 1987).

How and Why Rodents Make Great Damage in Alfalfa Fields

Over the growing season, rodents cause considerable yield losses in alfalfa crops by eating vegetative plant parts. Such losses are seasonal, or restricted to a part of the season, and it is possible to amend them during a succeeding period by timely activities, i.e., by taking the most appropriate measures of protection at the right time. Probably the most serious damage is caused by rodents gnawing at the underground parts of alfalfa plants, but also by digging their systems of burrows and thus disturbing the soil layer around plant roots. The possibility of root regeneration is minimal as such roots are open to infection by various soil pathogens, bacteria or fungi. Over the drier part of the year, plants with injured roots are unable to grow and begin to wither. This type of damage in alfalfa fields is constant and total, and it cannot be compensated for at a later time.

Due to their specific manner of living, their tunneling and type of nutrition, especially during outbreaks (to be discussed further down in this chapter), voles are considered potentially dangerous pests of alfalfa crops that are able to cause significant economic losses. When their population density reaches >200 animals/ha, it affects yield significantly, causing reductions or even total loss of yield (Babinska-Werka, 1979; Johnson, 1958; Ružić, 1983). A population of *M. pennsylvanicus* has been reported to cause 4 % yield loss in an alfalfa field (Johnson, 1958).

An average *M. arvalis* abundance of 145-285 animals/ha can cause 7.4-21.2 % alfalfa foliage yield decrease (Babinska-Werka, 1979; Jokić et al., 2010; Jokić et al., 2012). Potential losses in alfalfa crops caused by eating and other activities of *M. guentheri* mostly range from 16 % to 25 %. When protection measures are inadequate, losses may reach are much as 50 % (Wolf, 1977).

Very few reports are currently available on the harmfulness of ground squirrels and pocket gophers in alfalfa fields, which is conspicuously inconsistent with the many claims of their importance for alfalfa production (Cash, 2009; Orloff and Carlson, 1995; Witmer and Engeman, 2007). The very methods used to assess damage by rodents in alfalfa fields may be a possible explanation. Some of them are based only on sampling green foliage or counting injured or cut off tillers (Aplin et al., 2003) and cannot be applied to estimate the totality of damage caused by ground squirells or pocket gophers in alfalfa crops. Feeding and burrowing activities can cause yield losses, but may also damage the harvesting equipment (Whisson et al., 2002).

A reported average yield loss in an alfalfa study caused by *S. beldingi* feeding and other activities reached 46 % (Sauer, 1984; Whisson et al., 1999; Whisson et al., 2002). The maximum losses in some alfalfa fields was 75 % (Whisson et al., 2002). Marsh (1994) reported that the average *S. beldingi* population of 304 animals/ha caused 2006 kg/ha alfalfa yield reduction.

Pocket gophers are able to cause damage that reaches 50 % in alfalfa crop (Luce et al., 1981; Miller, 1946).

VOLE OUTBREAKS

Vole populations exhibit seasonal and multiannual variation, traditionally termed 'cycles'. Discussions about protection of various crops, alfalfa included, from vole damage cannot avoid mentioning this phenomenon, or its unpredictability and importance. Researchers have been trying for several decades to explain their causes, especially the cause of the spatio-temporal overpopulation (Krebs and Myers, 1974; Stenseth, 1977) but these events still remain erratic and unpredictible.

Based on our present knowledge, however, it is possible to select several factors that are believed to stimulate cyclic vole overpopulation (Blank et al., 2010; Carslake, 2011; Cockburn and Lidicker, 1983; Getz et al., 2006; Hestbeck, 1986; Huitu et al., 2005; Imholt et al., 2011; Klemola et al., 2002; Norrdahl and Korpimäki, 1996; Norrdahl and Korpimäki, 2002; Sundell et al., 2013):

- high survival of offspring, which is particularly significant for pups born during the vegetation season
- favorable weather conditions, e.g., warm weather during plant dormancy and optimal rainfall during vegetation
- uniform habitat occupancy, the quality and quantity of food that secures low spatial dissipation of animals
- topography and soil properties, e.g., loose soils are good for building tunnels
- small number of predators in the environment.

Generally, each factor contributing to the survival and reproduction of voles enhances the possibility of their overpopulation in a subsequent period. This is especially evident during years when several of these factors coincide in their activity.

According to our present knowledge, some factors are not believed to stimulate cyclic vole overpopulation (Boonstra, 1994; Getz et al., 2006; Hestbeck, 1986; Klemola et al., 2002):

- the number of females that are potentially reproductive has no crucial effect on vole outbreaks
- the density of previous vole overpopulation has no impact on a next outbreak
- nonuniform habitat occupancy; distinct spatial dissipation of animals increases the risk of predators and decreases survival expectancy
- low survival of offspring; this especially refers to pups born in late summer and early autumn
- the quality of individuals has no impact on cyclic density fluctations of voles
- age distribution; a prevalence of older members of a population and its aging process reduce the potential for overpopulation and rather contribute to decreasing the number of animals.

Differences among animal species, different climatic conditions and persistency of certain factors excluded from analyses are some of the reasons for having such divergance of data in various reports. For example, Krebs (1966) found that predators had no significant effect on the number of voles, i.e., that predators do not reduce their potential for overpopulation.

Another study (Boonstra, 1994) showed that physiological stress had an impact on animal numbers but Tkadlec and Zejda (1998) rejected it as a possible explanation of overpopulation on the grounds of simultaneous overpopulation events of several taxa taking place over large geographical areas.

As it is crucial to know the cause of vole overpupulation, especially the multiannual cyclic overpopulation, the phenomenon will certainly remain at the focus of research in different research disciplines. By combining several major factors that stimulate overpopulation and excluding those that do not, we strive to develop a multifactorial model for forecasting vole outreaks. In the context of economic impact of voles on plant production, including the profitable alfalfa production, it would be highly beneficial to be able to predict periods of outbreaks.

Data reported by a research group (Krebs et al., 2004), showing that up to 70 % success rate is possible in predicting house mouse outbreaks in South-Eastern Australia, is an incentive to continue research on vole outbreaks.

SOLVING THE PROBLEM

Apart from control measures applied in the surrounding area, the protection of alfalfa fields from rodent pests begins with pre-sowing cultivation. Deep tillage is especially important as it helps reduce the number of rodent animals. For example, it may downsize a common vole population by up to 70 % (Jacob, 2003).

Trapping, shooting or wire fences are useful methods for monitoring populations or reducing their size on a small area but these management strategies generally have low efficacy and poor cost-effectiveness on large alfalfa fields (Tickes, 1983), and deterrence of rodents in alfalfa fields by sound or vibration has also proved insufficiently effective (Baldwin and Holtz, 2010). Good choice of an alfalfa cultivar can partially alleviate the negative effect of rodents on alfalfa production in some regions (Case et al., 1997). Biological control includes the release of predators (owls, hawks, foxes, coyotes, snakes, etc.) into alfalfa fields or around them. Concerning owls, the choice of a location for installing nest boxes is very important for their settlement and reproducibility. Predator perches set up in alfalfa fields can stimulate the efficiency of hawks and eagles.

After planting a new field, it is necessary to maintain the number of rodent animals at an acceptable economic level.

In periods of the year when rodents may be expected to increase their numbers, and the risk of sustaining loss in alfalfa production is rising, effective protection measures need to be taken that would include toxic baits or fumigants. However, before using any toxicant, it is very important to consider closely the relevant legislation that sets conditions or restrictions for the use of rodenticides in alfalfa fields. The use of anticoagulant rodenticides, such as bromadiolone, may be restricted (Anonymous, 2011) or permitted (Janjić and Elezović, 2008; Vukša et al., 2012) in alfalfa crops. If chemical rodent control is not allowed in alfalfa crops at any time of the year, surrounding areas near field borders should be treated instead at any time.

A rodenticide registered for a particular purpose will not necessarily be effective (Jokić, 2012; Proulx et al., 2010; Tickes, 1983). Successfulness in controlling rodents does not depend solely on the best choice of a toxicant, its toxicity or the mode of action. Neither the application rate (per ha or active hole or burrow) nor the content of active ingredient in a bait will be a guarantee of success. For example, testing the efficacy of some acute and anticoagulant rodenticides, their different concentrations and formulations against pocket gophers, Tickers (1983) found a very low level of efficacy of no more than 18 %.

The method and rate of application, various agricultural and environmental conditions and alternative food sources may significantly reduce the efficacy of chemical rodent control. In alfalfa fields, rodenticides are mostly laid down manually in open holes or burrows, or applied mechanically, by broadcasting or by airplanes. However, the method and rate of bait application need to comply with the manufacturer's recommendation. Placing more bait in open active holes or burrows, or using higher application rates per area square unit would not promote successful control. On the contrary, harmful effects of rodenticides on nontarget organisms would only multiply (Brakes and Smith, 2005; Dowding et al., 2010). Optimization of rodenticide use heightens the protective effect while at the same time reducing possibilities of accidental poisoning of nontarget organisms. This is very important from the aspect of protection of biodiversity of native, as well as introduced vertebrate or invertebrate species (Elliott et al., 2013; Stojnić et al., 2011; Stojnić et al., 2013 in press).

The effectiveness of rodenticides is greatest when optimal amounts of bait are laid in active holes or burrows. Satisfactory results in rodent control on small areas can be achieved by laying baits in bait boxes. In order to get the best results, it is necessary to know the home range of rodent species and area effectively covered by each bait box (Kalinowski and deCalesta, 1981; Salmon et al., 2007). It is also important to protect bait boxes from other warm-blooded organisms that may happen to share the habitat (Whisson, 1997). The most cost-effective method of application is the spreading or laying of appropriate amounts of bait either mechanically or aerially. Those are the most frequently used methods of application of rodenticides in alfalfa fields, primarily of zinc phosphide. As its application is only allowed during alfalfa dormancy, when plant cover is very scarce, it increases the hazard for granivorous birds (Ramey and Sterner, 1995; Ramey et al., 2000).

Unfavorable agrometeorological conditions during rodent control may be one of the causes of rodenticide ineffectiveness (Jokić, 2012). That is particularly true of treatments with zinc phosphide as high humidity significantly reduces its efficacy.

As zinc phosphide has demonstrated a high level of effectiveness in controlling voles and squirrels in alfalfa fields (Table 1) its use in microencapsulated products would reduce the negative effect of environmental humidity and enable additional flexibility for treatment timing.

Table 1. Efficacy of some rodenticides and fumigants in rodent pest management control in alfalfa fields

Rodenticide or fumigant	Content of a.i. (%)	Baits type	Species	Efficacy (%)	References
Bromadiolone	0.005	Treated wheat grain	*Microtus arvalis*	78.08-85.12	Jokić et al., 2010.
Brodifacoum	0.005	Block		93.77-96.42	
Cholecalciferol	0.075	Granule		78.15-78.66	
Sodium selenite	0.1	Granule		67.52-67.73	
Cellulose	45.0	Pellets		82.75-87.51	
Bromadiolone	0.005	Treated wheat grain	*Microtus arvalis*	74.09-84.13	Jokić, 2012
Brodifacoum	0.005	Block		77.38-100	
Cholecalciferol	0.075	Granule		49.33-84.47	
Sodium selenite	0.1	Granule		50.91-75.51	
Cellulose	45.0	Pellets		22.07-37.50	
Bromadiolone	0.005	Treated wheat grain	*Microtus arvalis*	80.09-83.48	Jokić et al., 2012.
		Treated triticale grain		86.66-88.95	
Cholecalciferol	0.075	Treated wheat grain		75.46-81.13	
		Treated triticale grain		81.98-86.43	
Sodium selenite	0.1	Treated wheat grain		79.81-82.08	
		Treated triticale grain		76.74-86.09	
Zinc phosphide	2.0	Mixed with wheat and oil	*Microtus arvalis*	61.8	Aria et al., 2007.
			Nesocia indica	39.1-62.1	
Bromethalin	0.01	Block	*Microtus arvalis*	71.9	
			Nesocia indica	83.0-91.3	
Bromethalin	0.01	Pellet	*Microtus arvalis*	81.4	
			Nesocia indica	74.2-80.1	
Zinc phosphide	2.0	Rolled oat groats	*Microtus canicaudus*	94.0	Sterner et al., 1996.
Strychnine	0.35	Milo	*Geomys bursarius*	73.0	Ramey et al., 2002.
	0.75			100	
	1.30			90.0	
Chlorophacinone	0.005	Pellet	*Geomys bursarius*	100	Baroch and Poche, 1985.
			Thomomys talpoides	95.0	

Table 1. (Continued)

Rodenticide or fumigant	Content of a.i. (%)	Baits type	Species	Efficacy (%)	References
Gophacide	0.1	Grain	*Geomys bursarius*	85-98	Ward et al., 1967.
	0.1	Grain	*Thomomys talpoides*	92.0	
	0.1	Grain	*Thomomys umbrinus*	83-96	
Zinc phosphide	2.0	Treated milo	*Thomomys bottae*	40-45	Tickes et al., 1982.
	2.0	Pellets		20.0	
	1.78	Wheat		5-10	
Strychnine	0.35	Treated milo		5-15	
	0.5	Treated oats		5.0	
Diphacinone	0.005	-*		5-10	
Acrolein	-	Gas	*Thomomys talpoides*	59.0	Matschke and McCann, 1998.
Gas cartridge contains sodium nitrate and charcoal	-	Cartridge	*Thomomys talpoides*	7.1-30.8	Matschke et al., 1995.
Strychnine	0.37	Grains	*Thomomys talpoides*	8-42	Proulx, 1998.
	0.40	Treated oats		30-58	
Zinc phosphide	2.0	Treated oats		8-25	
Strychnine	0.5	Treated alfalfa hay	*Thomomys townsendi*	51.0	Lewis and O'Brien, 1986.
	1.0	Treated alfalfa hay		82.0	
	1.7	Treated alfalfa hay		75.0	
Strychnine	0.5	Paste-alfalfa bait	*Thomomys townsendi*	67.0	Matschke et al., 1993.
	1.5			71.0	
	3.0			80.0	
Aluminum phosphide	-	Tablets	*Spermophilus richardsonii*	36.0-87.6	Proulx et al., 2011.
Strychnine	0.05	Cabbage baits	*Spermophilus richardsonii*	83-88	Balliette et al., 2006.
Zinc phosphide	0.13		*Spermophilus mollis*	70-76	

Rodenticide or fumigant	Content of a.i.	Baits type	Species	Efficacy (%)	References
Aluminum phosphide	55	15-20 pellets	*Spermophilus beldingi*	94-98	Baldwin and Quinn, 2012.
Gas cartridges	-	Gas cartridges		100	
Acrolein	-	Gas	*Spermophilus beecheyi* *Spermophilus beldingi*	92.0	Clark, 1994.
Oxygen and propane gases	-	Gas	*Spermophilus beldingi*		Whisson et al., 2002.
Sodium monofluoroacetate	0.00625	Chopped cabbage bait	*Spermophilus beldingi* *Spermophilus townsendi*	92-97	O'Brien, 1978.
Strychnine	0.29			86.0	
Strychnine	0.19			66-81	
Zinc phosphide	0.8			39.0	
Aluminum phosphide	55	Tablets	*Spermophilus beecheyi*	96-100	Salmon et al., 1982.
Gas cartridges (sulfur, charcoal, red phosphorous, sodium nitrate)	-	Cartridges		66-72	
Aluminum phosphide	55	Tablets	*Spermophilus beecheyi*	100	Baldwin and Holtz, 2010.
Gas cartridges (carbon monoxide and carbon dioxide)	-	Cartridge		62-86	
Aluminum phosphide + gas cartridges	-	Tablets + Cartridge		59-71	
Gas cartridges	-	-	*Ground squirrel*	30-40	Orloff et al., 1995.

High moisture in the soil layer in which baits are laid, or rainfall, particularly over the first few days after treatment, stimulate mold development and bait degradation, reduce its persistence and result ultimately in unsuccessful rodent control (Jokić, 2012).

Sodium selenite and cellulose baits have generally shown changing levels of efficacy in the field (Jokić, 2012; Jokić et al., 2010; Jokić et al., 2012) and in control of commensal rodents (Đedović et al., 2011; Đedović et al., 2012). Adding aditives in baits with fungicidal or fungistatic properties, which have no negative impact on acceptance and palatability for rodents (Jokić et al., 2013; Marsh, 1988) is one of the ways to reduce the destructive effect of soil moisture.

The texture of soil, which is naturally porous, and its moisture bear upon the efficacy of fumigants. Allowing gas to move through soil, dry and porous soils have a negative effect on its efficacy. The best effects can be achieved by fumigation when soil is optimally moist (Whisson and Salmon, 2007). Additionally, the quality of alternative food sources can also affect significantly the effectiveness of rodenticides. Consumption of food that is rich in vitamin K may reduce the efficacy of some anticoagulants (Witmer et al., 2013).

Low efficacy can also be the consequence of incorrect assessment of the presence of a particular pest species or poor timing of toxicant treatment. Only by determining the rodent species correctly, catching animals of that species and correct assessment of active holes, colonies and characteristic signs of their presence (as explained) it is possible to choose the most appropriate measure for their control. This is also very important from the aspect of protection of rodent species that have a status of protected species. Some rodent species, such as hamster, *Cricetus cricetus*, can be found in alfalfa fields (Đedović et al., 2012). Even though they are able to cause considerable damage, their control is banned.

The best results in reducing the size of a vole population can be achieved by applying control measures during plant dormancy (Orloff and Carlson, 1995; Ružić, 1983) or just before the beginning of vegetation when ground squirrels and pocket gophers need to be controlled (Whisson and Salmon, 2007). The use of fumigants or zinc phosphide baits is not recommended for pocket gopher control due to low acceptability and effectiveness (Marsh, 1994; Tickers, 1983; Whisson and Salmon, 2007).

CONCLUSION

Profitable alfalfa production depends on whether rodents are present in the field and their rate, on available control methods and the achieved level of efficacy. Unfortunately, a consistent program of protection that would overcome the problem of losses from rodents in alfalfa crops has not yet been developed. The lack of a model that should coordinate:

- the biological properties of rodent species in the field
- meteorological parameters (primarily precipitation, irradiation, humidity and wind)
- micro-agroecological conditions existing in alfalfa crop (primarily moisture of top soil)
- type and texture of soil
- presence of nontarget organisms, and
- effects of nonlethal rodent control methods

with a possibility to choose the most effective method of control, i.e., rodenticides, fumigants and other toxic agents, in order to achieve the best control results is still an unsurmoutable obstacle in hitherto research and a challenge for the future.

As the present technology of rodenticide application has significant negative effects on nontarget organisms (Proulx, 2011; Ramey and Sterner, 1995; Ramey et al., 2000), it is important to offer new solutions, equally practical but more acceptable from the aspect of environment protection. Therefore, determining the level of improvement of rodent control in alfalfa fields by using microencapsulated zinc phosphide remains one of the main and immediate objectives.

REFERENCES

Anonymous, (2011). Review report for the active substance bromadiolone finalised in the Standing Committee on the Food Chain and Animal Health at its meeting on 11 March 2011 in view of the inclusion of bromadiolone in Annex I of Directive 91/414/EEC. European Commission Health & Consumers Directorate -General.

Aplin, K. P., Brown, P. R., Jacob, J., Krebs, C. J., & Singleton, G. R. (2003). *Field Methods for Rodent Studies in Asia and the Indo-Pacific*. Canberra, Australia: Australian Centre for International Agricultural Research.

Aria, A. K., Morovati, M., & Naseri, M. (2007). Efficacy of the Rodenticide Bromethalin in the Control of Microtus arvalis and Nesokia indica in Alfalfa Fields. *Turkish Journal of Zoology, 31,* 261-264.

Askham, L. R. (1994). Franklin, Richardson, Columbian, Washington and Towsend groun squirrels. *The Handbook: Prevention and Control of Wildlife Damage*. Lincoln, Nebraska: University of Nebraska.

Babinska-Werka, J. (1979). Effects of common voles on alfalfa crops. *Acta Theriologica, 24*, 281-297.

Baldwin, R. A., & Quinn, N. (2012). The Applicability of Burrow Fumigants for Controlling Belding's Ground Squirrels in Alfalfa. In R. M. Timm, & R. A. Baldwin (Eds.), *Proceedings of the 25th Vertebrate Pest Conference* (Pp. 160-163), Monterey, California: 25th Vertebrate Pest Conference (Mar 5-8, 2012).

Baldwin, R.A., & Holtz, B. A. (2010). Fumigation of California ground squirrels revisited: are fumigants aneffective methodfor controlling ground squirells? In R. M. Timm, & K. A. Fagerstone (Eds.), *Proceedings of the 24th Vertebrate Pest Conference* (Pp.129-132), Sacramento, California: 24th Vertebrate Pest Conference (Feb 22-25, 2010).

Balliette, J., O'Brien, J. M., & Eisemann, J. D. (2006). Efficacy of Strychnine and Zinc Phosphide Cabbage Baits in Controlling Ground Squirrels in Diamond Valley, Nevada. In R. M. Timm, & J. M. O'Brien (Eds.), *Proceedings of the 22nd Vertebrate Pest Conference* (Pp. 151-155), Berceley, California: 22nd Vertebrate Pest Conference (Mar 6-9, 2006).

Banfield, A. W. F. (1974). *The Mammals of Canada*. Toronto, Canada: National Museums of Canada, University of Toronto.

Banin, D., Abraham, H., & Zeev, A. (1993). Heat production in the Levant vole *Microtus guentheri*. *Israel Journal Of Zoology, 39,* 131-137.

Baroch, J., & Poche, R. M. (1985). Preliminary field evaluation of a new formulation of Rozol (chlorophacinone) bait against pocket gophers in Colorado. *In Great Plains Wildlife Damage Control Workshop Proceedings* (Pp. 138-144).

Batzli, G. O., & Pitelka, F. A. (1970). Influence of Meadow Mouse Populations on California Grassland. *Ecology, 51*, 1027-1039.

Benedix, J. H. (1994). A predictable pattern of daily activity by the pocket gopher *Geomys bursarius*. *Animal Behaviour, 48,* 501–509.

Blank, B. F., Jacob, J., Petri, A., & Esther, A. (2010). Topography and soil properties contribute to regional outbreak risk variability of common voles (*Microtus arvalis*). *Wildlife Research, 38,* 541-550.

Boonstra, R. (1994). Population cycles in microtines: the senescence hypothesis. *Evolutionary Ecology, 8,* 196-219.

Brakes, C. R., & Smith, R. H. (2005). Exposure of non-target small mammals to rodenticides: short-term effects, recovery and implications for secondary poisoning. *Journal of Applied Ecology, 42,* 118-128.

Carslake, D., Cornulier, T., Inchausti, P., & Bretagnolle, V. (2011). Spatio-temporal covariation in abundance between the cyclic common vole *Microtus arvalis* and other small mammal prey species. *Ecography, 34,* 327-335.

Case, R. M., Baker, D., Luchsinger, J., & Jasch, B. (1997). Cultural control damage to alfalfa caused by pocket gophers. In C. D. Lee, & S. E. Hygnstrom (Eds.), *Proceedings of the Thirteenth Great Plains Wildlife Damage Control Workshop* (Pp. 106).

Cash, D. (2009). *Alfalfa Management guide for Ningxia: Developing modern and sustainable alfalfa production systems in the Ningxia Hui Autonomous Region.* Beijing, P.R. of China: United Nations Food and Agriculture Organization.

Clark, J. P. (1994). Acrolein as a ground squirrel burrow fumigant. In W. S. Halverson, & A. C. Crabb (Eds.), *Proceedings of the 16th Vertebrate Pest Conference* (Pp. 78-79), Santa Clara, California: 16th Vertebrate Pest Conference (Feb 28 – Mar 3, 1994).

Cockburn, A., & Lidicker, W. Z. (1983). Microhabitat Heterogeneity and Population Ecology of an Herbivorous Rodent, *Microtus californicus. Oecologia, 59,* 167-177.

Cudworth, N. I., & Koprowski, J. L. (2010). *Microtus californicus* (Rodentia: Cricetidae). *Mammalian Species, 867/868,* 230-243.

Đedović, S., Vukša, M., Stojnić, B., & Jokić, G. (2011). Selenium Based Product and Sinantrop Rodents Control in Agricultural Facilities. *Pesticides and Phytomedicine, 26,* 265-270.

Đedović, S., Vukša,M., Petrović, M., Bojkovski, J., Pavlović, I., Jokić, G., & Stojnić, B. (2012). Control of brown rat (*Rattus norvegicus*) on a dairy farm in Serbia. *Biotechnology in Animal Hausbandry, 28,* 623-633.

Dowding, C. V., Shore, R. F., Worgan, A., Baker, P. J., & Harris, S. (2010). Accumulation of anticoagulant rodenticides in a non-target insectivore, the European hedgehog (*Erinaceus europaeus*). *Environmental pollution, 158,* 161-166.

Elliott, J. E., Hindmarch, S., Albert, C. A., Emery, J., Mineau, P., & Maisonneuve, F. (2013). Exposure pathways of anticoagulant rodenticides to nontarget wildlife. *Environmental Monitoring and Assessment,* (in press). pp. 1-12 DOI 10.1007/s10661-013-3422-x

Evans, F., & Holdenried, R. (1943). A population study of the Beechey ground squirrel in Central California. *Journal of Mammalogy, 24,* 231-260.

Fogel, R., & Trappe, J. M. (1978). Fungus Consumption (Mycophagy) by Small Animals. *Northwest Science, 52,* 1-31.

Gettinger, R. D. (1984). A field study of activity patterns of *Thomomys bottae. Journal of Mammalogy, 65,* 76–84.

Getz, L. L., Oli, M. K., Hofmann, J. E., & McGuire, B. (2006). Vole population fluctuations: Factors that initiate and determine intervals between them in *Microtus Pennsylvanicus. Journal of Mammalogy, 87,* 841-847.

Grzimek, B. (1990). *Grzimek's Encyclopedia of Mammals*: Volume Three. New York, N.Y.: McGraw-Hill Publishing Co.

Gulotta, E. F. (1971). *Meriones unguiculatus. Mammalian Species, 3,* 1-5.

Hestbeck, J. B. (1986). Multiple Regulation States in Populations of the California Vole, *Microtus Californicus. Ecological Monographs, 56,* 161-181.

Huitu, O., Laaksonen, J., Norrdahl, K., & Korpimäki, E. (2005). Spatial synchrony in vole population fluctuations - a field experiment. *Oikos, 109,* 583–593.

Imholt, C., Esther, A., Perner, J., & Jacob, J. (2011). Identification of weather parameters related to regional population outbreak risk of common voles (*Microtus arvalis*) in Eastern Germany. *Wildlife Research, 38,* 551-559.

Jackson, H. H. T. (1961). *Mammals of Wisconsin.* Wisconsin, Madison: University of Wisconsin.

Jacob, J. (2003). Short-term effects of farming practices on populations of common voles. *Agriculture, Ecosystems & Environment, 95,* 321–325.

Jacob, J., & Hempel, N. (2003). Effects of farming practices on spatial behaviour of common voles. *Journal of Ethology, 21,* 45–50.

Janjic, V., & Elezovic, I. (2008). *Pesticides in agriculture and forestry in Serbia*. Belgrade, Serbia: Plant Protection Society of Serbia.

Jenkins, S. H., & Eshelman, B. D. (1984). *Spermophilus Beldingi. Mammalian Species, 221*, 1-8.

Johnson, E. W. (1958). Consumption of alfalfa and wild oats by *Microtus californicus. Journal of Wildlife Management, 22,* 433-435.

Jokić, G. (2012). *Distribution, harmfulness and the possibility of rodents control in a field using different formulations of natural and synthetical rodenticides*. Phd thesis. Faculty of Agriculture, University of Belgrade, Belgrade, Serbia.

Jokić, G., Vukša, M., Đedović, S., Stojnić, B., Kataranovski, D., & Šćepović, T. (2013). Effects of Different Essential Oils on the Acceptability and Palatability of Cereal-Based Baits for Laboratory Mice. *Pesticides and Phytomedicine, 28,* 111-116.

Jokić, G., Vukša, M., Elezović, I., Đedović, S., & Kataranovski, D. (2012). Application of grain baits to control common vole *Microtus arvalis* (Pallas, 1778) in alfalfa crops, Serbia. *Archive of Biological Science, 64,* 629-637.

Jokić, G., Vukša, P., & Vukša, M. (2010). Comparative efficacy of conventional and new rodenticides against *Microtus arvalis* (Pallas, 1778) in wheat and alfalfa crops. *Crop Protection. 29,* 487-491.

Jones, C. A., & Baxter, C. N. (2004). *Thomomys bottae. Mammalian Species, 742,* 1-14.

Kalinowski, S. A., & DeCalesta, D. S. (1981). Baiting regimes for reducing ground squirrel damage to alfalfa. *Wildlife Society Bulletin, 9,* 268-272.

Klemola, T., Korpimäki, E., & Koivula, M. (2002). Rate of population change in voles from different phases of the population cycle. *Oikos, 96,* 291–298.

Krebs, C. J. (1966). Demographic changes in fluctuating populations of *Microtus californicus. Ecological Monographs, 36,* 239-273.

Krebs, C. J., & Myers, J. H. (1974). Population cycles in small mammals. *Advances in Ecological Research, 8,* 267-399.

Krebs, C. J., Kenney, A. J., Singleton, G. R., Mutze, G., Pech, R. P., Brown, P. R., & Davis, S. A. (2004). Can outbreaks of house mice in south-eastern Australia be predicted by weather models? *Wildlife Research, 31,* 465-474.

Lewis, S. R., & O'Brien, J. (1986). The Effectiveness of Strychnine Laced Alfalfa On Pocket Gopher Activity In Diamond Valley, Nevada. In T.P. Salmon (Ed.), *Proceedings of the 12th Vertebrate Pest Conference* (Pp. 116-119), San Diego, California: 12th Vertebrate Pest Conference (Mar 4-6, 1986).

Luce, D. G., Case, R. M., & Stubbendieck, J. L. (1981). Damage to alfalfa fields by plains pocket gopher. *Journal of Wildlife Management, 45,* 258–260.

MacClintock, D. (1970). *Squirrels of North America*. New York, New York: Van Nostrand Reinhold Company.

Mackin-Rogalska, R. (1981). Spatial structure of rodent populations co-occurring in different crop fields. *Polish Ecological Studies, 7,* 213–227.

MacMahon, J. A. (1999). Northern Pocket Gopher (*Thomomys talpoides*). In D. E. Wilson, and S. Ruff. (Eds.), *The Smithsonian book of North American mammals* (pp. 474-477). Washington, Washington: Smithsonian Institution Press in association with the American Society of Mammalogists.

Marsh R. E. (1994). *Belding's, California, and rock ground squirrel: The Handbook: Prevention and Control of Wildlife Damage.* Lincoln, Nebraska: University of Nebraska.

Marsh, R. E. (1988). Bait additives as a means of improving acceptance by rodents. *EPPO Bulletin, 18,* 195–202.

Marsh, R. E. (1998). Historical review of ground squirrel crop damage in California. *International Biodeterioration and Biodegradation, 42,* 93–99.

Maser, C. O., & Storm, R. M. (1970). A Key to Microtinae of the Pacific Northwest: Oregon, Washington, Idaho (1st Edition). Corvallis, Oregon: O.S.U. Bookstores, Inc.

Matschke, G. H., McCann, G. R., Krupa, H. W., & O'Brien, J. (1993). Field Evaluation of Three Strychnine Concentrations for Controlling Townsend's Pocket Gophers. *In Great Plains Wildlife Damage Control Workshop Proceedings* (Pp. 167-171).

Matschke, G. H., Ramey, C. A., McCann, G. R., & Engeman, R.M. (1995). Evaluation of a 2-active ingredient gas cartridge for controlling northern pocket gophers. *International Biodeterioration and Biodegradation, 36,* 151-160.

Matschke, G. H., & McCann, G. R. (1998). Evaluation of acrolein as a fumigant for controlling northern pocket gophers. In R. O. Baker, & A. C. Crabb (Eds.), *Proceedings of the 18th Vertebrate Pest Conference* (Pp.227-232), Costa Mesa, California: 18th Vertebrate Pest Conference (Mar 2-5, 1998).

Matthew, C. B. (2011). *Geomys bursarius* (Rodentia: Geomyidae). *Mammalian Species, 43,* 104-117

Mendelssohn, H. (1972). Ecological effects of chemical control of rodents and jackals in Israel. In M. T. Farvar, & Milton, J. P. (Eds.), *The Careless Technology: Ecology and International Development* (Pp. 527-544). New York, New York: Natural History Press.

Migula, P. (1969). Bioenergetics of pregnancy and lactation in the European common vole. *Acta Theriologica, 14,* 167-179.

Miller, M. A. (1946). Reproductive rates and cycles in the pocket gopher. *Journal of Mammalogy, 27,* 335–358.

Mueller, S. C., Undersander, D. J., & Putnam, H. D. (2007). Alfalfa for Industrial and Other Uses. In C. G. Summers and H. D. Putnam (Eds.), *Irrigated alfalfa Management in Mediterranean and Desert zones* (chapter 19, pp. 299-304). Oakland, CA: University of California and Natural Resources.

Norrdahl, K., & Korpimäki, E. (1996). Do nomadic avian predators synchronize population fluctuations of small mammals? a field experiment. *Oecologia, 107,* 478-483.

Norrdahl, K., & Korpimäki, E. (2002), Changes in population structure and reproduction during a 3-yr population cycle of voles. *Oikos, 96,* 331–345.

O'Brien, J. M. (1978). Chopped cabbage baits for ground squirrel control in Nevada. In W. E. Howard (Ed.), *Proceedings of the 8th Vertebrate Pest Conference* (Pp. 25-27), Sacramento, California: 8th Vertebrate Pest Conference (Mar 7-9, 1978).

Orloff, S. B., & Carlson, H. L. (1995). *Intermountain Alfalfa Management.* Oakland, CA: University of California, Division of Agriculture and Natural Resources.

Orloff, S. B., Salmon, T. P., & Gorenzel, W. P. (1995). Vertebrate Pests. Ch. 10 (Pp. 85-94) in : S. B. Orloff and H. L. Carlson (Eds.), *Intermountain alfalfa management.* University of California, Division of Agriculture and Natural Resources Publication 3366, Davis, California.

Ostfeld, S. R. (1985). Limiting Resources and Territoriality in Microtine Rodents. *The American Naturalist, 126,* 1-15.

Patton, J. L., & Brylski, V.P. (1987). Pocket Gophers in Alfalfa Fields: Causes and Consequences of Habitat Related Body Size Variation. *The American Naturalist, 130,* 493-506.

Proulx, G. (1998). Evaluation of Strychnine and Zinc Phosphide Baits to Control Northern Pocket Gophers (*Thomomys Talpoides*) in Alfalfa Fields in Alberta, Canada. *Crop Protection, 17,* 135-138.

Proulx, G., & Cole, J. P. (2002). Evidence of a second litter in Nothern Pocket Gophers, Thomomys talpoides. *Canadian Field-Naturalist, 116,* 322-323.

Proulx, G., MacKenzie, N., MacKenzie, K., & Walsh, K. (2011). Efficacy of aluminum phosphide tablets to control Richardson.s ground squirrel (*Spermophilus richardsonii*) populations in southern Saskatchewan, Canada. *Crop Protection, 30,* 1039-1042.

Proulx, G., MacKenzie, N., MacKenzie, K., Walsh, K., Proulx, B., & Stang, K. (2010). Strychnine for the control of Richardsons ground squirrels: efficiency and selectivity issues. In R. M. Timm, & K. A. Fagerstone (Eds.), *Proceedings of the 24th Vertebrate Pest Conference* (Pp.125-128), Sacramento, California: 24th Vertebrate Pest Conference (Feb 22-25, 2010).

Ramey, C. A., Matschke, G. J., Hegdal, P. L., McCann, G. R., & Engeman, R. M. (2002). Safe efficacy of three strychnine alkaloid bait concentrations for hand-baiting control of plains pocket gophers. *International Biodeterioration and Biodegradation, 49,* 139-143.

Ramey, C. A., & Sterner, R. T. (1995). Mortality of gallinaceous birds associated with 2% zinc phosphide baits for control of voles in alfalfa. *International Biodeterioration and Biodegradation, 36*, 51-64.

Ramey, C. A., Bourassa, J. B., & Brooks, J. E. (2000). Potential risks to ring-necked pheasants in California agricultural areas using zinc phosphide. *International Biodeterioration and Biodegradation, 45,* 223-230.

Ricankova, V., Sumbera, R., & Sedlacek, F. (2007). Familiarity and partner preferences in female common voles, *Microtus arvalis. Journal of Ethology, 25,* 95-98.

Rogers, L. M., & Gorman, M. L. (1995). The population dynamics of small mammals living in set-aside and surrounding semi-natural and crop land. *Journal of Zoology, 236,* 451-464.

Ružić, A. (1983). Rodents (Rodentia). In: *Manual for the Agricultural Monitoring and Forcasting Service.* (pp. 57-86). Belgrade, Yugoslavia: Association of Plant Protection Societies of Yugoslavia.

Salmon, T. P., Gorenzel, W. P. & Bentley, W. J. (1982). Aluminum phosphide (Phostoxin) as a burrow fumigant for ground squirrel control. In R. E. Marsh (Ed.), *Proceedings of the 10th Vertebrate Pest Conference* (Pp. 143-146), Monterey, California: 10th Vertebrate Pest Conference (Feb 23-25, 1982).

Salmon, T. P., Whisson, D. A., Berentsen, A. R., & Gorenzel, W. P. (2007). Comparison of 0.005% and 0.01% diphacinone and chlorophacinone baits for controlling California ground squirrels (*Spermophilus beecheyi*). *Wildlife Research, 34,* 14-18.

Sauer, W. C. (1984). Impact of the Belding's ground squirrel, *Spermophilus beldingi*, on alfalfa production in northeastern California. In D. O. Clark (Ed.), *Proceedings of the Eleventh Vertebrate Pest Conference* (Pp. 20-23), Sacramento, California: Eleventh Vertebrate Pest Conference (Mar 6-8, 1984).

Scheffer, T. H. (1948). *Mound Construction of the Pocket Gopher*. The Murrelet, 29, 9-10.

Sera, W. E., & Early, C. N. (2003). *Microtus montanus*. *Mamalian Species, 716,* 1-10.

Singleton, G. R., Hinds, L. A., Leirs, H., & Zhang, Z. (1999). *Ecologically-based Rodent Management*. Canberra, Australia: Australian Centre for International Agricultural Research.

Smith, M. F. (1998). Phylogenetic relationships and geographic structure in pocket gophers in the genus Thomomys. *Molecular Phylogenetics and Evolution, 9,* 1-14.

Stenseth, N. C. (1977). On the Importance of Heterogeneity for the Population Dynamics of Rodents: Towards a Theoretical Foundation of Rodent Control. *Oikos, 29,* 545-552.

Sterner, R. T., Ramey, C. A., Edge, W. D., Manning, T., Wolff, J.O., & Fagerstone, K. A. (1996). Efficacy of zinc phosphide baits to control voles in alfalfa - an enclosure study. *Crop Protection, 15,* 727-734.

Stojnić, B., Jokić, G., Đedović, S., & Vukša, M. (2013). Damage Caused by Vineyard Snail *Cernuella virgata* (Da Costa) to Green Onion Crop. *Pesticides and Phytomedicine* (in press).

Stojnić, B., Vukša, M., Jokić, G., & Čkrkić M. (2011). First Record of Introduced Valencia Slug, *Lehmannia valentiana* (Férussac, 1822), in Serbia. *Pesticides and Phytomedicine, 26,* 213-220.

Štrbac, P., Klokocar-Šmit, Z., Konstantinovic, B., & Dražic, D. (1996). *Zaštita lucerke od štetocina, bolesti i korova*. Novi Sad, Yugoslavia: Feljton. (In Serbian).

Sundell, J., O'Hara, R.B., Helle, P., Hellstedt, P., Henttonen, H., & Pietiäinen, H. (2013). Numerical response of small mustelids to vole abundance: delayed or not?. *Oikos, 122,* 1112–1120.

Thorne, D. H., & Andersen, D. C. (1990). Long-term soil-disturbance pattern by a pocket gopher, *Geomys bursarius*. *Journal of Mammalogy, 71,* 84–89.

Tickes, B. R., Cheath, L. K., & Stair, J. L. (1982). A comparison of selected rodenticide for the control of the common valley gopher (*Thomomys bottae*). In R. E. Marsh (Ed.), *Proceedings of the 10th Vertebrate Pest Conference* (Pp. 201-204), Monterey, California: 10th Vertebrate Pest Conference (Feb 23-25, 1982).

Tickes, B. R. (1983). Gopher control in alfalfa. *In Proceedings of the 13th California Alfalfa Symposium* (Pp. 68-71). Holtville, California: 13th California Alfalfa Symposium (Dec 7-8, 1983).

Tkadlec, E., & Zejda, J. (1998). Small rodent population fluctuations: the effects of age structure and seasonality. *Ecology and Evolution, 12,* 191-210.

Truszkowski, J. (1982). The impact of the common vole on the vegetation of agroecosystems. *Acta Theriologica, 27,* 305–345.

Verts, B. J., & Carraway, L. N. (1999). *Thomomys talpoides. Mammalian Species. 618,* 1-11.

Vukša, M., Jokić, G., & Đedović, S. (2009). IPM strategy for rodent control in alfalfa crops. *Biotechnology in Animal Husbandry, 25,* 1241-1248.

Ward, A. L., Hegdal, P. L., Richens, V. B., & Tietjen, H. P. (1967). Gophacide, a New Pocket Gopher Control Agent. *The Journal of Wildlife Management, 31,* 332-338.

Whisson, D. (1997) Vertebrate pest control in alfalfa hay. *In Proceedings of the 27th California Alfalfa Symposium* (Pp. 147-154). Visalia, California: 27th California Alfalfa Symposium (Dec 10-11, 1997).

Whisson, D. A., & Salmon, T.P. (2007). Integrated management of vertebrate pests in alfalfa. In C. G. Summers and H. D. Putnam (Eds.), *Irrigated alfalfa Management in Mediterranean and Desert zones.* (chapter 12, pp. 187-196). Oakland, CA: University of California and Natural Resources.

Whisson, D. A., Orloff, S. B., & Lancaster, D. L. (2002). The economics of managing Belding's ground squirrels in alfalfa in northeastern California. In L. Clark, J. Hone, J. A. Shivik, R. A. Watkins, K. C. Vercauteren, & J. K. Yoder (Eds.), *Proceedings of the Third NWRC Special Symposium* (Pp. 104-108), Fort Collins, Colorado: Third NWRC Special Symposium (Avg 1-3, 2000).

Whisson, D. A., Orloff, S. B., & Lancaster, D. L. (1999). Alfalfa yield loss from Belding's ground squirrels in northeastern California. *Wildlife Society Bulletin, 27,* 178-183.

Witmer, G. W., & Engeman, R. M. (2007). Subterranean Rodents as Pests: The Case of the Pocket Gopher. In S. Begall, H. Burda, & C. E. Schleich (Eds.), *Subterranean Rodents: News from Underground* (pp. 287-299), Heidelberg - Berlin, Germany: Springer-Verlag.

Witmer, G. W., & Singleton, G. R. (2010). Sustained agriculture: the need to manage rodent damage. In F Wager, (Ed.), *Agricultural production* (chapter 1, pp. 1-38). New York, New York: Nova Science Publications.

Witmer, G. W., Snow, N. P., & Moulton, R. S. (2013). The effects of vitamin K1-rich plant foods on the efficacy of the anticoagulant rodenticides chlorophacinone and diphacinone, used against Montane Voles (*Microtus montanus*). *International Journal of Pest Management, 59,* 205-210.

Wolf, Y. (1977). The Levant vole, *Microtus guentheri* (Danford et Alston, 1882). Economic importance and control. *EPPO Bulletin, 7*, 277–281.

Yavuz, M., Öz, M., & Albayrak, I. (2010). Levant vole (*Microtus guentheri*) (Danford and Alston 1880) prefers southerly-facing slopes in agricultural sites at Antalya, Turkey. *North-Western Journal of Zoology, 6,* 36-46.

Zimmerman, E. G. (1999). Plains Pocket Gopher (*Geomys bursarius*). In D. E. Wilson, and S. Ruff. (Eds.), *The Smithsonian book of North American mammals.* (pp. 485-486). Washington, Washington: Smithsonian Institution Press in association with the American Society of Mammalogists.

INDEX

A

B

C

D

E

F

G

H

S

T

U

V

W

Y

Z